INTEGRATED APPROACH TO RURAL DEVELOPMENT

About the Author

G.N. Karalay is presently a guest faculty at National Power Training Institute, Nagpur. After graduating in Mechanical Engineering in 1962, he served various government organisations like C.E.A., BHEL and NTPC from where he retired as Executive Director. Later he served Power Finance Corporation as its Senior Consultant. Also, while working in several mega green field projects he was closely associated with resettlement and rehabilitation of rural people, who were displaced as a result of land acquisition.

INTEGRATED APPROACH
TO
RURAL DEVELOPMENT
Policies, Programmes and Strategies

G.N. KARALAY
B.E.(Hons.) Mech. FIE
Former Executive Director, NTPC and
Sr. Consultant, Power Finance Corporation

CONCEPT PUBLISHING COMPANY, NEW DELHI-110059

ISBN-13: 978-81-8069-225-3 (HB)
ISBN-13: 978-93-5125-216-0 (PB)

Edition 2016

Published and Printed by

Concept Since 1974

Concept Publishing Company Pvt. Ltd.
A/15-16, Commercial Block, Mohan Garden
New Delhi-110059 (India)
T: 25351460, 25351794, *F*: 091-11-25357109
E: publishing@conceptpub.com, W: www.conceptpub.com
Editorial Office: H-13, Bali Nagar, New Delhi-110 015, India.

Cataloging in Publication Data--Courtesy: D.K. Agencies (P) Ltd. <docinfo@dkagencies.com>

Karalay, G. N., 1939- author.
Integrated approach to rural development : policies, programmes and strategies / G.N. Karalay.
pages cm
Includes bibliographical references and index.
ISBN 9788180692253 (HB)
ISBN 9789351252160 (PB)

1. Rural development--Government policy--India. 2. Rural development projects--India. 3. Sociology, Rural--India. I. Title.

HN690.Z9C6 2016 DDC 307.14120954 23

to

the fond memory

of

my beloved parents and aunt,

"Nani"

Preface

The population in the rural area constitutes 72.22 percent of the total population of India. Although the progress since Independence in the area of Rural Development is not small, the goals set before the country in various fieids of Rural Development like basic education for all, primary healthcare for all and development in agriculture and allied areas of Rural Development have remained unaccomplished although target dates for the same have been getting revised from time to time.

Rural Development is a complex phenomenon and it calls for Integrated Approach that combines action on many fronts. Basic education, primary healthcare, development of agriculture, infrastructure and efficient delivery system are essential for Rural Development Besides, no programme can be undertaken without adequate financial resources. India as a still developing country has its own limitations in this respect Although higher investment in all the fields of Rural Development is definitely called for, the problem does not end there and proper utilisation of the scarce financial resources becomes an important component. Similarly no worthwhile development can take place without commitment of political class, accountability of officials and active participation of rural community. Thus it is a matrix of multiple factors that determines the success of any programme taken in hand. So far the focus has been on a single factor without giving required attention to the other factors. Recognising the limitations of isolationist approach, this book has propounded in its place, integrated approach to Rural Development. Based on my case studies and personal field experience, it is found that in order to achieve the integrated approach, certain strategies including institutional changes are required to be implemented. These strategies form prominent part of this book.

The book is divided in six chapters namely Chapter 1: Rural Sociology— Socio Economic Structure of Rural Society; Chapter 2: Basic Education in Rural Area: A Long Journey Ahead; Chapter 3:

Agriculture and Rural Development; Chapter 4: Primary Healthcare in Rural Area—A Perspective; Chapter 5: Trend Analysis of Budget Allocations to Social and Rural Sectors; and Chapter 6: Matrix of the Ingredients of Integrated Approach to Rural Development.

The uniqueness of this book lies in the extensive coverage of basic education, primary healthcare and various aspects of rural development in line with the three components forming part of the UN Human Development Index, which is the only accepted measure of Nation's progress at the International level.

I hope that this book will be of immense use to the policy makers, planners, project implementers, social organisations, research workers, academicians, students taking up competitive examinations and public in knowing the background of the important socio-economic issues and the vital changes required to be brought in the socio-economic scenario of the rural India.

I thank my wife, Kunda Karalay, for assisting me in every way and bearing with me for my working everyday for long hours in completing this work and my children for giving me good moral support to work for this research venture.

I thank profusely my colleague Dr K.K. Govil for offering most valuable suggestions and highly encouraging me in publishing this book. I thank my colleague Rajeshwar Datt for assisting me by way of providing some of the latest books on the subject matter. My sincere thanks are also to Shri Ashok Kumar Mittal and his dedicated team in Concept Publishing Company for bringing out this book in such a nice form and within a short period of time.

G.N. KARALAY

Contents

List of Tables

List of Figures

Introduction

The development of any country is basically synonymous with the development of the people living in that country. Rural population which by and large is underserved constitutes 72.22 per cent (2001 Census) or nearly three fourth of India's total population. Therefore, for the country's overall development, all out efforts are required to be targeted towards development of rural area. Human Development Index prepared by the United Nations Development Programme is so far accepted as a most proper measure of the progress of any country. In terms of Human Development Index which constitutes three basic components viz. literacy, life expectancy and income, India ranks as poorer as 127 out of 177 countries. In spite of larger population, China has a better ranking. It is obvious that lack of adequate progress in providing these three basic components especially in rural area have attributed to India's low rank in the comity of nations.

After Independence India has adopted democratic form of governance and rightly so. There are certain rights given in the Constitution to citizens about providing basic facilities in rural area. Our founding fathers have provided a Constitution with directive principles, if followed in all seriousness, should have led us to become a developed country by now. Even China with larger population and without the framework of democracy is ahead of us in many ways. The peculiarity of China is that it has persisted with high emphasis on basic education, primary healthcare and rural development. This gives us a lead to examine our position with respect to these basics.

Since the time India became independent, it has witnessed some measure of economic growth. However, the measure of success of the economy should not be only just how much quantity of assets or profits are generated but it should be that after generation how equitably these ace distributed. Further, while sustained economic growth will tend to reduce overall poverty, education and healthcare are the most important assets of the poor as through only these they will be empowered to get proper employment and status in the society.

There has been emphasis in five-year plans on providing to all, three basic components i.e. basic education, primary healthcare and a level of development that will ensure at least minimum standard of living in rural area. The idea has been to improve the capability of India's underserved population so that they can actively participate in the Nation's development process. India is rich in natural and human resources. But there are glaring disparities in our country. India is surplus in food and we are exporting food but millions in our country are remaining hungry, because our public distribution system has not functioned properly. In high technology areas like Space and Information Technology we are counted among seven leading countries in the world but we have almost the lowest literacy rate. People from Europe are coming all the way to our country for specialist health treatment but there is no access even to the primary healthcare in large part of the rural area.

Society has several needs and each section of the society is engaged in providing a particular need for example agriculturist is providing food and fibre, businessman is providing a product and engineer is providing a facility to make the life easier. These and several of other sections of the society are working in tandem on different aspects of life. Since what is being contributed by the rural population comes under essential category, he is entitled for equivalent return if not better for his output with respect to others. But the situation as exists in the society today is entirely different. Average agriculturist today is standing almost at the lowest rung of the ladder in terms of the prosperity and status in the society.

As it is long since we achieved our independence, this is an appropriate time to revisit the progress made in the field of Rural Development and what needs to be done to achieve the goals set before the nation.

Many of the schemes initiated by the Govt. though good in intent necessitate detailed planning as well as proper approach to implementation. Rural masses are generally not fully aware of these schemes. There are greater chances of success if the people for whom these schemes are meant are someway involved in planning and implementation processes.

Further, allocation of funds for various fields of rural development is far too inadequate taking into account the actual requirement. This is evident from steadily decreasing investments in agriculture in terms of percentage of GDP. The delivery of the service in terms of quality is another area of concern.

At the macro level, the success depends upon commitment of the political class for the development process, accountability of the officials for planning and implementation and wholehearted participation, in the development process, of the community for whose benefit these schemes are designed. The Panchayat Raj and cooperative institutions which are vital for bringing the development in the rural area need to be strengthened properly. Also the progress in all the fields of rural development depends upon development of proper infrastructure, allocation of adequate financial resources and efficient delivery system. In nutshell only an integrated approach to the whole development process has a chance to succeed.

In this book attempt has been made to integrate the above components so that it becomes easier to implement the schemes successfully. The book is divided in six chapters namely Chapter 1: Rural Sociology: Socio Economic Structure of Rural Society; Chapter 2: Basic Education in Rural Area—A Long Journey Ahead; Chapter 3: Agriculture and Rural Development; Chapter 4: Primary Healthcare in Rural Area—A Perspective; Chapter 5: Trend Analysis of Budget Allocations to Social and Rural Sectors; and Chapter 6: Matrix of the Ingredients of Integrated Approach to Rural Development. Each part chronologically consists of historical background; constitutional provisions; various approaches adopted so far; policies/programmes under implementation; present status and constraints; and strategies to be adopted to achieve the goal of All Round Rural Development.

1

Rural Sociology: Socio Economic Structure of Rural Society

1. INTRODUCTION

1. Importance of Rural Development

Although there has been steady increase in the process of urbanisation, out of the country's total population of 1028.7 (*2001 census*) million, 743 million (i.e. 72.22 per cent of total population) still live in villages. Village truly represents our cultural heritage. Therefore what Gandhiji said long ago," India lives in villages and village constitutes the very heart of India" holds true even today. Agriculture is the mainstay of the country as 65-70 per cent of the country's population is dependent on agriculture for their livelihood and its produce cater to the basic necessities of the life of the people. However the rural people still suffer from poor living conditions. 26 per cent of the country's population or 267.5 million falls Below the Poverty Line. Out of this, the rural area where BPL percentage is 27.1 constitutes 201.4 million population. Internationally the poverty is measured with income below $1 per day and $2 per day, in terms of purchasing power parity. On this basis, as per UNDP HDR 2004, the percentage of poor in India is 34.7 and 79.9 respectively. Accordingly India accounts for 357 million and rural area 258 million poor even on the basis of income of $1 per day. In spite of overall economic growth of the country as a whole, there is no significant change in the living conditions of the rural population, on the other hand the disparities between the urban and rural population have increased tremendously. It is against this background that Rural Development assumes great importance.

2. Definition and Objectives of Rural Development

Dictionary meaning of development is growth or evolution, stage of advancement. In the context of rural background it means developing better physical, social and economic conditions of the population living in the rural area. The definitions and objectives of the rural development as laid down by the authors vary according to their area of emphasis. All the same, the objectives have been better reflected in the views of Michael Todaro that Rural development encompasses: (i) Improvement in levels of living, including employment, education, health and nutrition, housing and a variety of social services; (ii) Decreasing inequality in the distribution of rural incomes and in rural-urban balances in income and economic opportunities, and (iii) Increasing the capacity of the rural sector to sustain and accelerate the pace of these improvements." The foregoing views are inclusive of all aspects that have bearing on rural development in true sense.

3. Defining Rural-Urban Area

Broadly, rural population is involved in the field activities for producing food, fibre and raw materials, whereas urban population is involved in the activities of processing the aforesaid items along with carrying on the industrial activity. Apart from occupational differences, the traits of the two societies differ in many ways like:

Rural population is scattered over a large area, whereas urban area is thickly populated. Rural area still lacks basic civic amenities, whereas in urban area there is an organised effort for creation of these facilities. Apart from these two broad based categories, there is a third category called *Rurban* which represents the intermediate scenario like those living in farms or places nearby the towns where they carry on dairy, poultry or such businesses. To make a clear distinction between these categories and collecting their census data has always posed a difficulty. However, Census in India in 1981 tried to identify the urban area as defined below, thereby meaning that the area not covered in this is a rural area.

(a) All statutory towns, i.e. all places with a municipal corporation, municipal board, Cantonment board or notified town area, etc.
(b) All other places which satisfy the following criteria:

(i) a minimum population of 5000;
(ii) 75 per cent of the male working population engaged in non-agricultural (and allied) activities; and
(iii) a density of population of at least 400 per sq km (or 1000 per sq mile)

2. RURAL SOCIOLOGY

1. Introduction to Rural Sociology

Before we proceed to discuss about rural development in depth, it is necessary to have insight of the rural sociology as it gives us a clue about the methodology to be adopted for achieving the goals of development in rural area. Dictionary meaning of sociology is study of society and social problems. It can further be defined as study of patterns which emerge out of the relations among the people. In the same way rural sociology means study of rural society and their social problems. Whatever has been stated under rural sociological aspects in this book is based upon study of number of villages in general and a village Natala (in Maharashtra) in particular with which the author has had close association throughout life.

India means rural India because nearly three-fourth of the country's population live here. It is in this context that rural sociology assumes great importance. Village is the basic unit of rural area. It represents humane approach and simplicity. Historically it also represented as nearest to the nature i.e. land, hills, forests, rivers, springs and on the human front, natural (but rich in itself) life, folk and folk-tales. Its richness has inspired great poets and writers. Peace and harmony in the hearts of the village people have been quite apparent

However beneath this peace and harmony there have also been deeper problems. These problems have basically been social as well as economic. The social problems emanated from discrimination on the basis of caste and economic problems from abject poverty in the lower strata of people. The rural society has been characterised as highly static society. The people in rural area were disinclined to move out and they have been trying to cope up with the problems without leaving the native place. In a way this was also due to absence of transport and communication facilities at that time. Agriculture and allied activities have been the main occupation of the rural population and as such life of the rural society moved around the calendar of the activities of the agriculture. For example religious functions, marriage ceremonies, festivals, family functions etc were all related to the agricultural activities.

Till the advent of British Rule the villages were functioning as autonomous units. The jurisdiction of the village heads extended to houses, streets, bazaars, wells, tanks etc. The village heads also settled disputes between the villagers. They organised community works and collected the taxes. After the advent of British Rule the basic structure of administration within the village was not materially changed. The local executive officer named *police-patil* was usually hereditary choice and not the member of the Govt.

2. Social Structure of Rural Society

2.1. Religion

Rural society on the national scale though predominantly Hindu has population belonging to all religions. At the outskirts of villages there are temples or shrines of village deities. They are symbols of deep religious faith. People in rural area follow their religion more ardently than their urban counterparts. Even where there are people belonging to more than one religion, in rural area, people have good interpersonal relationship and live in harmony. The functions in a family belonging to a particular religion are also attended by the people of other religions wholeheartedly. True spirit of brotherhood is visible in rural area. That is why even when there is turmoil between two religious groups in towns and cities, the villages are not at all affected.

2.2. Caste System

The Indian civilization is evolved out of pre-existing folk and regional cultures. It is rich with traditions of great epics-Ramayana, Mahabharata and other religious scriptures. This great tradition maintained its continuity through regions, villages, castes and tribes. The origin of our society is in village. We come to see even today all facets of old social relationships existing there, though may be with slight degree of variation. Social structure in the village is based on caste system as a lineage of the olden time four *varnas* namely *Bramhin, Kshatriya, Vaishya and Shudra.* However in our ancient civilization, the system of *varnas* was different from the caste system as started being practiced later. Originally the four-fold division of labour for functioning of society has been prescribed in the *Vedas* and *Bhagvad Gita.* In Chapter-4 Verse-13 Sri Krishna says, "The four *Varnas* (order of society) were created by me according to the *gunas* (attributes) and actions of men)" Sri Krishna further says in Chapter-18 Verse 41, "The duties of *varnas* have been divided according to their *gunas* (attributes)" God never differentiates human beings on the basis of caste, colour or race, which is evident from the sunshine and elements that sustain life as well as earthquake (on 26.12.2004) unleashing Tsunami that kills more than 1,50,000 human beings within a few hours. *Nowhere in Gita, varna by birth is stated.* This interpretation is given by Paramahansa Yoganandji in his epic, "God Talks to Arjuna-The Bhagavad Gita" which is one of the most outstanding works of a great sage. In this context, he states "In material sense all nations recognize the natural divisions of castes, according to ability: viz., the labourers, the businessmen, the soldiers and rulers, the teachers and clergymen. A labourer can become a businessman or a clergyman; therefore these four groups are intechangeable on the material plane." Except for the purpose of duty, there was no distinction in social matters between the *varnas.* However over a period of time due to development of vested interests by the privileged, aberration in the form of birth determining the *varna* or the caste crept in. This is how we now find the caste system has come to stay.

Untouchability which has its origin in caste system is a further aberration. It is an evil of serious nature; The scheduled caste people suffered from serious discrimination due to the practice of untouchability. They were not allowed to enter the temple or draw water from the wells along with the other caste people. It is due to the sincere efforts of Mahatma Gandhi, Dr B.R. Ambedkar and other social reformers that the society has undergone social transformation and the evil practice in its strongest form has lost much of its following. However the practice is still prevalent in soft form in some parts of the country and even now active participation of scheduled castes in development schemes is not significantly visible.

There is a hierarchical gradation of castes from highest to the lowest. *Brahmin* caste group and persons owning large area of land were considered to be the highest in the hierarchy. There is however now a reflection of political influence on this hierarchy and presently that caste is dominant which has more numerical strength and exercises more influence economically and politically.

2.3. Cultural Mobility

The caste system thrived on traditional customs. Although the attempt has been to maintain basic characteristic unchanged, there has been attempt in one caste to adopt some good features of the other caste, which has been referred as a process of *Sanskritisation*. The present economic scenario has only accelerated this cultural mobility.

2.4. Joint Family System

Traditionally there has been a *joint family system* in vogue in India. However with passage of time and industrialisation in which people have to leave homes, in urban area the joint family system is giving way to a nuclear family system. Such a transformation has however not taken place in rural area in any significant manner. It is because primarily the values in rural people are still based on intense personal relationship within the family and secondly this system also helps them to look after their agriculture without hiring persons from outside the family. There are however instances when there are divisions in the family after the marriage of the member of the family thus leading to fragmentation of land holdings.

2.5. Institution of Marriage

The castes are further divided in sub-castes reflecting the division of social relationship at micro level. Every person becomes a member of the caste by birth and thus caste, marriage and other customs are highly interrelated. Marriage is an important social institution. Its purpose goes beyond fulfilling biological need. In urban area, although caste system does play important role in establishing marriage relationship, there is a fair amount of mobility towards inter-caste marriages. The education and relationship at the work place have started in some measure influencing the marriage relationship. However in rural area there is no such change evident and people here are still guided by caste and sub-caste when marriage relationship is to be established. Apart from choosing marriage relationship, other traditional customs are still followed the same way as they used to be. There is thus a strong bond of relationship within the caste. However such relationship does not exclude the social relationship with otter caste people in the village. For example, when there is any marriage ceremony or occurrence of death in a family, all people in the village join in sharing the joys or bereavement in that family. The participation by the community in the functions, unlike in urban area, is not just symbolic but they provide physical and other necessary help to the family.

2.6. Jajmani System

There are several relationships which cut across the caste system. *Jajmani system* is one such. This is based on patron-client relationship. On occasions like birth, marriage and death, Jajmans perform special duties for which they are paid a customary sum of money or some gifts in kind. Occupationally Darji, Nai, Dhobi, Kumbhar, Carpenter, Silversmith are some of the castes who give their services to their clients on exchange basis. The Jajmani system also went beyond the concept of single village unit to that encompassing number of villages. However due to mass manufactured cheap products available in the market, the preponderance of Jajmani system in a village is on way out. At least the practice of return in kind as exchange basis is coming to an end. However there are some occupational services which are given on cash payment and they would continue to remain so.

In a broader sense Jajmani system of patron-client relationship is taking a different shape in the form of institutional arrangement like cooperatives covering number of villages.

2.7. Village Unity

Village is largely a self-sufficient unit. It has a capability to produce for its peoples' minimum needs everything except say salt or metal. Village functions practically as a self-governing unit. The characteristic of identification to the community is another factor for unity. It is pertinent to observe that on the occasion of birth, death, marriage or religious functions, there is a wide participation of villagers cutting across caste or other distinctions. Agriculture is the main occupation of the rural people. There are three major social groups having different status. These are big landlords, small farmers and landless labourers. Although, there is a vertical division of the society on the lines of caste and ownership of land, factors like living in a small community and almost everybody being engaged in agriculture, have resulted in a very high level of interdependency, cutting across caste or economic divisions. The rural people therefore though may belong to different caste groups or may be small or marginal farmer, a common thread runs through binding all villagers together.

It presents a unique multirelationship matrix. The group of elders in village play important role in resolution of the conflicts whether these are inter-caste or intra-caste and they are largely successful in this task. The collective consciousness in the villagers is so strong that it has endured many challenges in the form of droughts, famines, epidemics or even wars.

3. SOCIAL STRATIFICATION IN RURAL AREA

The dictionary meaning of stratification is arrangement in strata or layers. The basis of social stratification is inequality. The inequality in class, status and power is found in any society. Referring purely to social stratification in rural area, it has broadly two dimensions, one that is based on caste and the other based on the amount of land owned. The stratification on the basis of caste has already been discussed earlier. As regards the other dimension, the number of villagers owning large area of land is very small due to land ceiling act as well as divisions in the family from generation to generation. Therefore the village stratification follows the grouping of villagers into small and marginal farmers and landless labourers.

4. SOCIAL TRANSFORMATION

There existed in past stratification on the basis of caste and the amount of land owned. The caste has always played important role in the village life and acted as a strong unifying factor within the caste. However there are other factors like interdependency between the villagers for day-to-day needs and heavy dependency of the villagers on the nature (like rain) for the production from the agriculture which made the position of the village to function as a micro-cosm single cohesive unit which goes beyond the caste.

The spread of literacy is gradually bringing changes in thinking as well as way of life. There is now relatively greater awareness in the rural society about their rights and responsibilities. Rural society which used to be a closed one, limiting their activities to one's own village, is getting aware of the opportunities outside their village and now more and more educated rural youth and landless labourers are seeking employment in towns and cities. The institutional development like formation of co-operative societies covering many villages together has also brought changes on the agrarian front. Land reforms have also impacted the rural society significantly.

The establishment of Panchayat Raj Institution (PRI) has further changed the mode of governance in the village. The well-to-do farmers have found expression in occupying the positions of political power. So also average villagers started getting recognition for their voting power. With the participation of rural community in the working of PRI, it has filled the gap which used to exist in the past between the village community and the outside world. The scheduled castes and scheduled tribes and women who were always out of power structure got represented in the structured village institutions through reservations. The major change in the rural sociology has thus been brought in by the development of inter-caste relationship, formation of the economic institutions like co-operative societies and establishment of new power structure i.e. Panchayat Raj Institutions. The entire pre-existed social structure is undergoing radical changes and new power structure dominated by political and economic power is emerging. Thus even though there still exists a bond of caste and land, rural society is undergoing slowly but steadily a phenomena of social transformation.

5. AGRARIAN SOCIAL STRUCTURE

Since ages India has been agriculture oriented country and even now 65-70 per cent of total population is dependent on agriculture for their livelihood. In view of this land has acquired great significance. For a villager land means more than a source of sustenance. He is not only economically but emotionally also attached to it. The dictionary meaning of "agrarian" is that which is related with land and its distribution. The agrarian system thus centers around management of land and its utilisation for productive purposes.

During pre-Independence period, British rule did not want to disturb the land ownership system existing then. Their main concern was collection of taxes and land revenue. After the First World War was over, the agrarian unrest assumed larger dimension. The Royal Commission on Agriculture was therefore appointed in 1928 to make recommendations for the improvement of agriculture. However the issues like making recommendations regarding the existing system of land ownership and tenancy or of assessment of land revenue and irrigation charges were kept outside the scope of its terms of reference.

During British Rule there existed systems like: (i) *Ryotwari*—under this tenancy system, every registered holder of land was recognised as its proprietor. He paid revenue directly to the Govt.

(ii) *Mahalwari*—under this system, land was held jointly by co-sharing bodies of village communities, (iii) *Jamindari/Malgujari*—under this system, one or more persons owned a village and were responsible for payment of the revenue, (iv) *Jagirdari*—under this system, jagirs were granted to certain military commanders, ministers and courtiers. Thus except the class of Ryotwari, the other classes acted as intermediaries. Therefore the first task of land reforms which were to be undertaken after Independence was towards abolition of intermediaries.

The purpose of land reform goes beyond looking land only as a source of livelihood. Land also symbolises social status. If a person owns a piece of land, even if it is a small one, it gives him a feeling of better social status in the village. It thus acts as a catalytic agent for empowering the rural people. The land reforms further include the aspect of registration of tenancy arrangements and regularise the land tenure records.

6. AGRARIAN MOVEMENTS

The discontent in the agriculturists has been a familiar feature in the past but it took a shape of organised struggle in 20th century. As mentioned earlier, at the end of First World War, the simmering discontent in the agrarian society reached the high point. The beginning of the agrarian crisis was accompanied by the entry of the agriculturists in the political field. When Gandhiji returned from South Africa, he offered his guidance to the agrarian movements in Champaran (Bihar) during 1917-18, in Kheda (Gujarat) in 1918 and Bardoli (Gujarat) in 1920. Other agrarian movements were Moplah Rebellion in Malabar in 1921, peasant revolt in Punjab in 1930, Telangana peasant struggle in Andhra Pradesh during 1947-51. The mass movements started by Gandhiji were essentially based on the larger rural question which included the land tax, security of the land tenure and exploitation of the poor. He also wanted to experiment through these movements his idea of non-cooperation which was found to be quite effective in South Africa. These circumstances compelled British Raj to take up the study of the problems of the villages although they also had in mind to use this study to assess the revenue potential of the villagers. Gradually the movement for Independence became so intense that the Govt. gave priority to maintenance of law and order rather than taking up agrarian reforms. Thus economic condition of the agriculturists right up to time of Independence did not improve. On the other hand it deteriorated due to famines, floods, epidemics, etc. The problems were further accentuated by general indebtedness in the poor on the large scale. As there were no commensurate measures from the Govt., either preventive or post occurrence, to mitigate these problems, the unrest in the agricultural community continued unabated till Independence.

7. OCCUPATIONAL STRUCTURE

There are three major modes of occupations in rural area. These are:

1. Agriculture and its Interrelated Activities

Most of the rural population (more than 70%) are dependent for their living on agriculture and its interrelated activities such as horticulture, cattle rearing, dairying, poultry, forestry, fishing, etc. The income generated from agriculture alone is generally not sufficient enough to bear the expenses of the family. Further there are some lean periods in a year when there is not much work in the agriculture. In view of this, many villagers are taking up allied activities to supplement the income from agriculture. On account of various credit schemes available, there is a shift in the occupational mode from that of agriculture alone to activities allied to agriculture.

2. Small Scale and Cottage Industries

For the same reasons as mentioned above, those in rural area who are educated and have entrepreneurial skill are taking to small scale and cottage industries. Small scale industries include soap making, oil *ghani,* etc. Cottage industries include producing khadi cloth and clothes, handloom, coir, carpets, sericulture, potteries, baskets, toys, etc. According to survey (2001-02) total employment in this sector amounted to 55.15 million. Out of this bulk of the employment has been in small scale industry (19.2 million), handloom industry (12.4 million) followed by *khadi and village industries* (6.01 million). This constitutes 13.2 per cent of the eligible working population in rural area. This is a welcome sign from the point of view of the need of growth in rural economy although this is not a significant figure compared to the overall need of the employment in rural area.

3 Services

The services sector include shopkeepers who sell provisions of daily use, middlemen who buy the produce at cheaper rate from the needy poor farmers and sell in towns at higher rates; tractor or thresher owners who give their equipment on hire; transport operators, etc. To improve economic status, the traditional service providers like carpenters, goldsmiths, artisans, cobblers etc. are gradually leaving these professions in the villages and migrating to nearby towns/cities where there are better opportunities.

8. POPULATION SHIFT

As per Census in India in 2001, rural population is 743 million which accounts for 72.22 per cent of the total population of the country. Table.1.1 shows the total population and its break up in rural and urban population. Over a period of time there is a steady declining trend in the rural population and correspondingly increasing trend in the urban population.

Table 1.1: Rural and Urban Population 1901-2001

Census Year	*AAEGR*	*Population (Million)*			*Percentage of Total Population*	
		Total	*Rural*	*Urban*	*Rural*	*Urban*
1901	–	239	213	26	89.2	10.8
1911	0.56	252	226	26	89.7	10.3
1921	(–) 0.03	251	223	28	88.8	11.2
1931	1.04	279	246	33	88.0	12.0
1941	1.33	319	275	44	86.1	13.9
1951	1.25	361	299	62	82.7	17.3
1961	1.96	439	360	79	82.0	18.0
1971	2.20	548	439	109	80.1	19.9
1981	2.22	683	524	159	76.7	23.3
1991	2.14	847	629	218	74.3	25.7
2001	1.95	1029	743	286	72.22	27.78

AAEGR = Average annual exponential growth rate (%)

Source: India 2005 (GOI Publication)

The difference in the living conditions and the availability of opportunities for betterment of future has an important bearing on the pace of shift in population from rural to urban area.

9. CONCLUSION

As mentioned above, because of lack of scope in villages for professional services coupled with educational base, there is a flight of youth from rural to urban area where there are more opportunities for the professional services as well as employment. Caste is thus gradually loosing its traditional influence on the choice of occupation. Due to educational attainments and occupational mobility, joint family system is also gradually breaking. With time, the rural society is thus undergoing social transformation. Notwithstanding efforts made in the past, most of the villages still face the same old problems as narrated in the chapters to follow in this book although to a lesser degree and rural people feel themselves as a neglected lot when they compare with better facilities of living and entertainment available in towns and cities. It is therefore clearly evident that efforts made in the past in the field of rural development are far short of the genuine expectations of the rural population.

2

Basic Education in Rural Area: A Long Journey Ahead

1. STATUS OF LITERACY AND CONSTITUTIONAL GUARANTEES

1. Importance of Education

The dictionary meaning of education is a process of teaching, training, and learning especially in schools and colleges to improve knowledge and develop skills. There is a strong relationship between development of the mankind and education. Without education, development can neither be broad based nor sustained. The child is born already with basic faculties but these faculties are not so well developed. The objective of education is to draw out the best in the child so that in course of time he becomes a useful member of the society. Towards this end, the main emphasis of education has to be on building the character of the child and developing his understanding of ethical values on one hand and acquiring knowledge, skill and competence on the other. Thus it has to be a blend of all these aspects. The days of our working in isolation are over and in 21st century we are living in a global scenario. The attitude of being harmoniously together is therefore to be developed from the early stage of the child so that he is well prepared for living in pluralistic and global situation when he is grown up. Through education, it is possible to enable an individual to grow as well informed confident, competent and responsible citizen and newly acquired capability will enable him to participate more effectively in the process of development.

2(a). Human Development Index (HDI)

How do we measure real progress of the country? This question has been engaging the attention of various world organizations since it was felt that either Gross National Product (GNP) or Gross Domestic Product (GDP) alone could no longer be considered a valid measure of progress as these reflect mainly economic aspects and therefore not enough to reflect the status of the progress in a comprehensive way. In this context in early 1990s United Nations Development Programme (UNDP) brought out *Human Development Index* (HDI) as a valid measure of the country's progress. It encompasses three major components. The education due to its unique importance is one of these components, others being longevity (indicator of health) and per capita Gross Domestic Product. Thus Human Development Index is calculated as an average of the sum of three indexes namely longevity, educational attainment and income. Individual Index is calculated from the equation:

$$\text{Index} = \frac{\text{Actual Value} - \text{Minimum Value}}{\text{Maximum Value} - \text{Minimum Value}}$$

Minimum and maximum values are fixed for each indicator.

1. Longevity is represented by the indicator-life expectancy at birth The minimum value for the indicator is 25 and maximum 85
2. Educational attainment is represented by a combination of the indicators of adult literacy rate with 2/3rd weight and gross enrolment ratio (i.e. combined primary, secondary and tertiary enrolment ratio) with 1/3rd weight. For each indicator, the minimum value is zero and maximum 100.
3. Income is represented by real GDP per capita in purchasing power parity US dollars. The minimum value for the indicator is US$ 100 and maximum US$ 40,000.

Table 2.1 shows the comparison between India and China as regards human development indicators. As per UNDP Human Development Report 2004, during 2002, India ranks 127 out of the total 177 countries in terms of Human Development Index as against the China's rank of 94. Since 72.22 per cent of India's population lives in rural area, the relatively less progress in the fields of basic education, primary healthcare and rural development is mainly responsible for the low ranking of India in terms of HDI.

Table 2.1: Human Development Indicators-Comparison between India and China 2002

Description	*India*	*China*
Combined primary, secondary and tertiary gross enrolment ratio (%)	55.0	68.0
Adult literacy rate (% age 15 and above)	61.3	90.9
Education index	0.57	0.79
Life expectancy at birth (years)	63.7	70.9
Life expectancy index	0.64	0.76
GDP per capita (PPP US$)	2670.0	4580.0
GDP index	0.56	0.62
Human development index (HDI) Value	0.595	0.745
Rank	127	94

Source: UNDP, Human Development Report 2004.

2(b). Millennium Development Goals

Emanating from the United Nations Millennium Declaration-2002, Millennium Development Goals have been formulated which are:

Goal 1: Eradicate extreme poverty and hunger.

Target 1: Halve, between 1990 and 2015, the proportion of people whose income is less than $1 a day (existing proportion is 34.7%).

Target 2: Halve, between 1990 and 2015, the proportion of people who suffer from hunger (existing undernourished people are 21%).

Goal 2: Achieve universal primary education

Target 3: Ensure that by 2015, children everywhere, boys and girls alike, will be able to complete a full course of primary schooling (existing net primary enrolement ratio is 83%)

Goal 3: Promote gender equality and empower women.

Target 4: Eliminate gender disparity in primary and secondary education, preferably by 2005 (existing net primary enrolment ratio of female to male is 0.83).

Goal 4: Reduce child mortality

Target 5: Reduce by two-thirds, between 1990 and 2015, the under five motality rate (existing under-five mortality rate is 93 per 1,000 live births).

Goal 5: Improve maternal health.

Target 6: Reduce by three-quarters, between 1990 and 2015, the maternal mortality ratio (existing MMR is 540 per 1,00,000 live births).

Note: 1. Figures in parenthesis pertain to India

Source: UNDP HDR 2004.

3. Present Status of Literacy in India

The term literacy rate is used for the population relating to seven years and above. A person is deemed to be literate if he or she can read and write with understanding in any language. A person who can merely read but cannot write is not considered literate. The literacy rate taking into account the total population in the denominator has now been termed as "crude literacy rate".

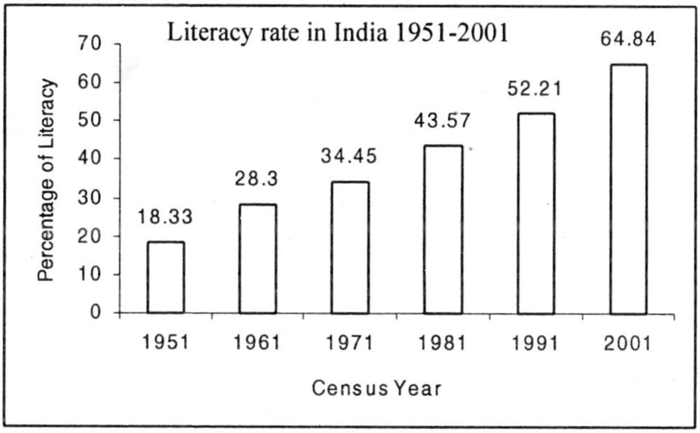

Source: Census of India

Fig. 2.1: Decadal Progress of Literacy Rate from 1951 to 2001

Note: Literacy rates for 1951, 1961, 1971 censuses relate to population aged 5 years and above. The rates for 1981, 1991 and 2001 censuses relate to population aged 7 years and above.

India is a vast country; it has an area of 32,87,263 sq km; 5,93,643 inhabited villages; 5161 towns and population of 1028.7 million (Census 2001). 72.22 per cent of this population lives in rural area. At the time of Independence, overall literacy rate was 18 per cent and female literacy was barely 9 per cent. The gross enrolment ratio at primary level (classes I-V with the age group of 6-11) was 43 per cent and the corresponding percentage for girls was only 25. At upper primary level (classes VI-VIII with the age group 11-14) enrolment ratio was only 12.5 per cent and the corresponding percentage for girls was barely 5.

As per *2001 Census*, the overall literacy rate in our country has reached 64.84 per cent. The significant milestone reached in India is that the total number of illiterates has come down from 328.88 million in 1991 census to 296 million in 2001 Census. Therefore for the first time, the decade has seen a decline in the absolute number of illiterates in the country. However as per the World Education Report, India still accounts for world's 33 per cent illiterates. In India the gross enrolment ratio has risen to about 80 per cent. However this figure does not take into account the children who have been enrolled but not actually attending the school. Such children are either engaged for providing help to the family at home, working in the agriculture or assisting parents in their vocation. According to one survey such "out of school children' are around 15 per cent. Thus while there is no doubt that the progress achieved so far is not small, there is still a long way to achieve 100 per cent literacy. The literacy rate in rural area is much less than urban area. Table 2.3. Shows literacy rate in India (urban-rural) for the years 1991 to 2000 as per 55th Round of National Sample Survey.

The literacy rate in 2000 in rural area has been 60 per cent as against 81 per cent in urban area. The gap between urban and rural literacy rates has however reduced from 28 per cent in 1991 to 21 per cent in 2000.

The gross literacy rate of 64.84 per cent (2001 Census) comprises the male literacy rate of 75.26 per cent and female literacy rate of 53.67 per cent. Traditionally people in general did not encourage girl child to get educated. It is borne out of the old notion that women have a primary role of raising the family and in any other area if they work it should be a supportive to the men's role. This view has worked to the disadvantage of the girl children and resulted in perpetuation of many injustices in the Society. In simple terms the best could not be drawn from half the population and society has become poorer to that extent. Those who were at the forefront of the freedom movement tried to bring reforms in all the social fields especially education. Considerable social transformation has taken place since then and a change in the outlook is quite discernible. There is now more awareness among larger section of both literate and illiterate parents about the importance of education of their children, irrespective of those being sons or daughters. Poorest of the poor also are showing willingness to send their children to schools even though it may mean parting with their meager resources. There has never been such upsurge in the society before in favour of educating the children. It is also surprising that rural women feel more strongly about educating the children than their male counterparts.

Gandhiji is one of those who played important role in bringing various social reforms.

Table 2.3: Literacy Rate in India (Urban-Rural) 55th Round (July 1999-June 2000)

State/Union Territory	*Rural*			*Urban*		
	Male	*Female*	*Persons*	*Male*	*Female*	*Persons*
All India						
1991	58	31	45	81	64.	73
1993	62	36	49	83	67	76
1994	63	36	50	85	68	77
1995	64	38	51	86	70	78
1996	65	39	52	86	71	79
1997	68	43	56	88	72	80
1998	69	44	57	87	72	80
1999	68	43	56	87	72	80
2000	72	47	60	88	73	81
State: 1998						
Andhra Pradesh	58	37	48	84	68	76
Arunachal Pradesh	62	47	55	92	71	82
Assam	85	69	78	95	89	92
Bihar	57	28	44	80	55	68
Goa	87	78	82	98	93	96
Gujarat	78	50	64	90	75	83
Haryana	72	45	59	89	69	81
Himachal Pradesh	81	63	71	96	86	92
Jammu & Kashmir	70	42	57	82	54	69
Karnataka	67	44	56	84	71	77
Kerala	94	88	91	97	93	95
Madhya Pradesh	62	34	49	83	66	75
Maharashtra	77	53	65	92	82	87
Manipur	86	71	79	89	81	85
Meghalaya	88	81	85	96	94	95
Mizoram	92	89	91	99	98	99
Nagaland	90	79	85	85	78	82
Orissa	64	41	53	89	74	82
Punjab	70	57	64	83	77	80
Rajasthan	62	24	44	88	66	78
Sikkim	83	64	74	83	84	83
Tamil Nadu	76	55	65	90	78	84
Tripura	88	75	82	95	88	92
Uttar Pradesh	65	35	50	80	58	70
West Bengal	77	56	67	90	79	85

Source: Statistical Abstract, India, 2003 (GOI Publication)

4. Gandhiji on Education

Gandhiji held the view that education does not mean merely a knowledge of letters but it means character-building, a knowledge of duty. Cleanliness of the mind and body is the first step in education. It should be built on a solid foundation of truth and purity of life and enable us to discriminate between good and bad, to assimilate the one and eschew the other.

According to Gandhiji, primary education deserved greatest importance. Such primary education would include, besides normal study, the elementary principles of sanitation, hygiene, nutrition, doing one's own work, helping parents at home, etc. The emphasis should be on education through manual training and by teaching a useful vocation. According to his conception, primary education should encompass the level of Matriculation and primary education should be free and compulsory.

The system of education expounded by Gandhiji known as *Nai Talim* included blending of theoretical studies with ethical values and vocational training. Although we have moved away from this concept, the incorporation of vocational training seems to be more relevant today in the context of paucity of employment opportunities for the youth coming out of schools and colleges.

5. Constitutional Position

The root cause of most of the socio-economic problems is widespread illiteracy and therefore this area calls for most urgent attention. At the time India attained Independence, the position of literacy was simply pathetic. As mentioned earlier, the overall literacy rate was 18 per cent and female literacy rate was only 9 per cent. The gross enrolment ratio at primary stage was 43 per cent and the same for girls was only 25 per cent. At upper-primary stage the enrolment ratio was barely 12 per cent and the same for girls was 5 per cent. Therefore how long will it take to attain a goal of education to all was a question before the Constituent Assembly at the time of drafting the Constitution. They had before them a report of the Sargeant Commission, constituted before Independence, suggesting a 40 year period for this. It was then thought that the country cannot afford to wait this long for so important a matter like providing education to all and thus Article 45 in this connection was introduced in the

Constitution. This article under the title—Provision of free and compulsory education for children- reads, "The State shall endeavour to provide, within a period of ten years from the commencement of this Constitution, for free and compulsory education for all children until they complete the age of fourteen years." While urgency was felt and rightly so at the time of framing a Constitution for providing the education in quickest possible time, total complexity of the problem was not realized at that time. What has followed subsequently was opening of more and more schools and the real goal remained far from realization. Realizing the stupendous task involved, the first comprehensive exercise was undertaken by the Education Commission during 1964-66. The Commission felt that the Universal Elementary Education (UEE) would require a period of 20 years, thereby meaning the period ending by the year 1985. The recommendations of the Commission included framework necessary to achieve the goal of UEE, adoption of non-formal education and establishing part-time schools. Looking to the slow pace of progress being achieved in increasing the outreach of the literacy, it was thought that the Central Govt. should directly come into the picture to step up the efforts of the States. It is in this context that the Constitution was amended in 1976 bringing the education in the Concurrent List. This has resulted in new way of sharing of responsibility between the Central Govt. and the States with regard to all aspects relating to education. The greatest benefit of this change was meant to enhance the pace of literacy by way of increasing the outreach of basic education.

6. Making Education a Fundamental Right

A global conference was held at Jomtien in 1990 under the auspices of UNESCO. The theme of the conference was Education for All and in this context to review the status of education in different parts of the world. The decade that followed Jomtien Conference witnessed many innovative schemes undertaken by the states in India. These schemes included adoption of non-formal education, establishing part-time schools and employing *Shiksha Karmis* (para-teachers). The decade also witnessed an important development in the form of Supreme Court judgment which, interpreting the constitutional provisions, stated basic education is a fundamental right requiring the State to make necessary provisions as a basic obligation. Making education a Fundamental Right in the age group of 6 to 14 years is an important step in meeting the popular aspirations of the people. To achieve this, the State will have to extend its reach to every village howsoever small and remote it is. Taking into account vastness of the country it would require great effort and enormous amount of resources.

The real difficulty lies in covering boys and girls from the rural area as schools are not available in small villages and children have to trod long distances. For girl children it becomes all the more difficult. Socio-economic conditions have also large bearing on the literacy levels. Besides, to increase the outreach of education, far larger allocation of financial resources to education, of the order of 6 per cent of GDP is called for as against 3.74 per cent (BE) and 3.0 per cent (RE) in 2003-04. The expenditure includes provision for salaries of the teaching staff, school building, teaching aids, textbooks, mid-day meals, sports items, regular health check up etc.

7. Free and Compulsory Education to Children Bill 2003

On account of its importance, providing Free and Compulsory Primary Education finds important place in the Directive Principles of the State Policy. The Parliament has passed the Constitution 86th Amendment Act, 2002 to make elementary education a Fundamental Right for children in the age group of 6-14 years. The important features of the mechanism to implement this include a teacher for every class room, teachers will have to ensure attendance, all elite schools will have to admit 20 per cent children from needy class etc. There are 8.5 lakh primary upper primary schools with only 33 lakh teachers. It is therefore herculean task to provide a teacher for every class. However the same bill provides for, besides approved schools, Educational Guarantee Centers (EGC) and alternative schools run under the approved scheme of central and state governments. In order to address the problem of arranging qualified teachers for EGCs and alternative schools, the bill provides for appointment of teachers who have passed class X examination or furthermore. These teachers will have to undergo 30-day training programme. It requires to be seen if such under qualified, under trained and underpaid teachers will be able to do justice to teaching as the curriculum in these schools will be the same as in approved schools.

8. Responsibility Towards Children Beyond 14 Years

According to the concept of Gandhiji's primary education should encompass the level of Matriculation i.e. up to the age of 16. Presently age group of 6 to 11 and 12 to 14 are only targeted for close monitoring. However this coverage will have to be extended up to the age of 18 years as defined in the Child Rights Charter to which our country is a signatory. As such policy framework has to reflect convergence of early childhood care, education and healthcare.

2. POLICIES AND PROGRAMMES

1. National Population Education Project (NPEP – School Education)

The population education assumes great importance in India against the backdrop of India being second largest populated country in the world, The National Population Education Project (NPEP-School Education) was therefore started in April 1980 with a view to make population education as part of the school education system. Earlier funded by the United Nation's Population Fund, it will now be funded by the Central Govt.

2. National Policy on Education (NPE)

The change of education from State list to Concurrent List did improve the situation as the Central Govt. stepped in augmenting the efforts of the states. The role of the states still remained pivotal. Depending on the local situation and initiative, many states started different schemes to increase the enrolment ratio. The Central Govt. continued to play a lead role in monitoring of these schemes. Based on detailed inputs from the states and prolonged nationwide debate on the problems confronting the education system prevailing then, the National Policy on Education (NPE) was launched in 1986 along with Programme of Action (POA) to achieve Universalising Elementary Education. The global conference held at Jomtien in 1990 set the goal of Education for All which became the guiding force for all countries especially for India. The next decade witnessed many schemes undertaken in India as a follow up of the education policy. Programme of Action was updated in 1992 and the modified policy now envisages, a National Approach to a system of education, bringing about uniformity in education, making adult education programme a mass movement, providing universal access, retention and quality in elementary education, special emphasis on education of girls etc.

3. Operation Blackboard (OB)

The schools generally faced the problem of non-availability of requisite number of teachers and even minimum facilities required. It was against this back ground that the scheme of Operation Blackboard was started in 1987-88. It was aimed at augmenting the strength of the teachers and physical resources available in primary schools. The scheme included providing two reasonably large rooms, at least two teachers and essential teaching-learning material for every primary school. During 1993-94, the scheme was extended to cover upper primary schools. The scheme included providing three rooms for such schools, an additional teacher for upper primary school and a third teacher for primary school with enrolment of more than 100 children.

The funds for operating the scheme were provided to State Governments by the Central Govt. The scheme has been subsumed in Sarva Shiksha Abhiyan from 2002-03.

4. Universalisation of Elementary Education (UEE)

The education is one of the three components constituting Human Development Index and thus progress in the field of education has assumed great importance. The concept of Universalisation of Elementary Education is synonymous with that of Education for All. The objective of UEE is to enlarge the outreach of education so as to cover every child of the age of 6 years and above under this programme. The task is so gigantic that it has been found difficult to achieve the goal with the pace of progress of formal education and availability of regular teachers. Therefore innovative modes like opening non-formal education centres and appointing para-teachers (*Shiksha Karmis*) have been employed. The results of this system are found to be varying depending upon the performance of the para-teachers appointed for the purpose.

Further, greater public involvement is a basic ingredient to make such innovative scheme successful. In this context, since it was found to be quite difficult to manage the schools by State alone, assigning this function to village committees has been found to be a good idea. This indicates that if there has to be decentralisation in any field, decentralisation in the field of education tops the list.

5. National Literacy Mission (NLM)/Adult Education (AE)

5.1. Objective of NLM

There is significant illiteracy especially in rural area in the age group of 15-35 years, which is the productive and reproductive age group and constitutes a major segment of the working population. The National Literacy Mission was thus founded in May 1988 with the aim of attaining a sustainable threshold level of 75 per cent literacy by 2007 by imparting functional literacy to this group. Besides pre-determined capability levels of reading, writing and numeracy with comprehension, functional literacy includes imbibing values of national integration, conservation of environment, women's equality, observance of small family norms, etc. Since most of the persons in the above age group are occupied with some or the other assignment during daytime, the Adult Education programme has to be conducted at a time which is suitable to them.

The Central Directorate of Adult education provides academic and technical resource support to the National Literacy Mission.

5.2. Total Literacy Campaign (TLC)

The Total Literacy Campaign (TLC) is the principal strategy of NLM to spread the literacy. The highlights of these campaigns are that these are area and time period specific, participative and objective oriented. These are implemented through Zilla Saksharta Samities (district level literacy committees), constituted as autonomous bodies. The Samities have due representation of various sections of Society. The campaigns are characterized by large scale mobilisation through mass based communication strategy. The survey undertaken at the grass-root level also serves as a tool of planning, mobilisation and environment building. Apart from imparting functional literacy, campaign also disseminates other socially important messages like enrolment and retention of children in schools, immunisation, propagation of small family norms, women's empowerment, communal harmony, etc. These campaigns in turn usually generate a demand for primary education.

5.3. Continuing Education

The basic literacy skills need nurturing. There is every possibility of neo-literates regressing into the position they started unless special efforts are made to consolidate, sustain and if possible enhance the literacy levels they have attained.

The Continuing Education Scheme provides a learning continuum to the efforts of Total Literacy and Post Literacy Programmes. The main thrust is on providing further learning opportunities to neo-literates through Continuing Education Centres (CECs) which provide area-specific, need based opportunities for basic literacy, upgradation of literacy skills, alternative educational programmes, vocational skills and also promote social and occupational development. The programmes of Residual Illiteracy are also being taken up to address the needs of geographically remote regions and segments of population requiring special focus particularly on SCs/STs/Women. The scheme also undertakes a number of important programmes such as: Equivalency Programme designed as an alternative education programme equivalent to the existing formal, general or vocational education; Income Generating Programme facilitating the participants to acquire or upgrade their vocational skills and take up income generating activities; Quality of Life Improvement Programme to equip learner and the community with essential knowledge, attitude, values and skills to raise their standards of living; and Individual Interest Promotion Programme providing opportunities for learners to participate and learn about their individually chosen social, health, physical, cultural and artistic interests.

6. District Primary Education Programme (DPEP)

Post global conference on Education for All held at Jomitien in 1990 saw some of the developed countries offering assistance to India in furthering improvement in literacy. The assistance is basically aimed at supplementing the resources of the State, which are not adequate enough to cater to the needs of the primary education to all. District Primary Education Programme is one such externally funded programme. In this programme local communities are encouraged to participate in improving the learning in the schools. The DPEP is aimed at providing additionality and is designed to fill in the gaps in other existing schemes by providing inputs over and above the provisions made under Central and State schemes for primary education.

The DPEP adopts a holistic approach to universalise access, retention and improve learning and to reduce disparities among social groups. It is intended to strengthen the capacity of national, state and district institutions existing in the field of primary education.

The programme includes construction of classrooms and new school buildings, opening of Non-Formal/Alternative Education Centres, appointment of new teachers, setting up early childhood education centers, strengthening of State Councils of Educational Research and Training (SCERTs)/District Institutes of Educational Training (DIETs), setting up of Block Resource Centres, teacher training, development of teaching-learning material, research based interventions, special interventions for promoting education of disadvantaged groups, girls, SC/ST, etc., initiatives for providing integrated education to disabled children and distance education for teacher training. The expenditure limits for civil work is 33.3 per cent, management cost is 6 per cent and the balance amount is earmarked for quality improvement activities.

The programme essentially is aimed at (i) reducing differences in the rate of enrolment, dropout and learning achievement between gender and social groups to less than 5 per cent, (ii) reducing overall primary dropout rates for all students to less than 10 per cent, (iii) raising average achievement levels by at least 25 per cent over measured baseline levels by ensuring achievement of basic literacy and numeracy competencies and minimum of 40 per cent achievement levels in other competencies by all primary school children, and (iv) providing access for all children to primary schooling or its equivalent non-formal education.

The districts for the programme are selected on the basis of: (i) educationally backward districts with female literacy below the national average, and (ii) districts where Total Literacy Campaigns (TLC) have been successful, leading to enhanced demand for elementary education.

DPEP has so far opened more than 1,60,000 new schools, including 84,000 alternative schooling (AS) centres. Besides there are centres covering bridge courses. Village Education Committees/ School Management Committees have been set up in villages/ habitations/schools covered by DPEP.

As already mentioned, DPEP is an externally aided project. 85 per cent of the project cost is met by the Central Govt. (through external aid) and remaining 15 per cent shared by the State Govt. The programme during last 9 years has covered almost half of the country.

7. National Programme of Nutritional Support to Primary Education

Abject poverty is the major reason for lack of enthusiasm in rural poor to send their children to school. Due to poverty, parents are compelled to engage their children to help them in their family avocation When they are unable to feed their children how will they send them to school? The National Programme of Nutritional Support to Primary Education commonly known as Mid-Day Meal Scheme was therefore started in August 1995. The programme is aimed to cover all students in primary classes in Govt., Govt. aided and local body schools. The objective of the scheme is to give a thrust to primary education by way of increasing enrolment, attendance and retention simultaneously impacting on (i.e. creating demand for enhancing) nutritional status of children in primary classes. The programme now extends to the children studying in EGS and other alternative learning centers. The support from the Central Govt. includes: (i) 100 gms of foodgrains per child per school day where there is a meal programme, alternatively 3 kg per child per month for 10 months, and (ii) admissible transportation charges. Foodgrains are supplied through FCI.

Looking to the aspect that some states did not start the Mid-Day Meal scheme, the Apex court directed all States which have not started providing cooked meals to children to introduce the same in a time bound manner. While some of the states have implemented the scheme fully by way of providing the meals, it is implemented partially in others. In some parts of the states only foodgrains are being distributed instead of cooked meals. The overall percentage of the children covered so far with cooked meals under the scheme is 35 per cent, coverage varying from state to state. However it is noteworthy that the scheme has been implemented in southern states more vigorously than rest of the states.

Based on the experience of the NGO-ISKCON, Karnataka Govt. sought to involve more NGOs in running the scheme. The foodgrains and transportation cost are made available to the NGOs and these NGOs take further responsibility of providing the cooked meals. Nandi Foundation in Hyderabad has organized a central kitchen which caters to as many as two lakh children. The scheme is in operation in most of the states with varying degree of progress.

8. Sarva Shiksha Abhiyan (SSA)

By the end of the decade of 1990's, it was found that the pace of improvement in literacy rate was not adequate enough to achieve the goal of UEE in the near future. Table 2.4 shows decadal enrolment ratio of children in primary and upper primary stages from 1950-51 to 1997-98.

Table 2.4: Enrolment ratio in Primary and Upper Primary Stages 1950-51 to 1997-98

Year	*Primary Stage (Classes I-V)*			*Upper Primary Stage (Classes VI-VIII)*		
	Boys	*Girls*	*Total*	*Boys*	*Girls*	*Total*
1950-51	60.8	24.9	42.6	20.8	4.3	12.9
1960-61	82.6	41.4	62.4	33.2	11.3	22.5
1970-71	96.5	60.5	78.6	46.3	19.4	33.4
1980-81	95.8	64.1	78.6	54.3	28.6	41.9
1990-91	113.9	85.5	100.1	76.6	47.0	62.1
1997-98	97.5	81.2	89.7	66.5	49.5	58.5

Source: DoE, MHRD, GOI/IER-2002

In 1997-98, the enrolment ratio of 89.7 per cent was achieved in primary stage and 58.5 per cent in upper primary stage. The ratio of enrolment for the same year in respect of girls was about 9 per cent less than the boys in both the stages. Another problem prominently faced was high dropout rate "in both primary as well as upper primary stages. Table2. 5 shows dropout rate in both the stages from 1960-61 to 1997-98.

Table 2.5: Dropout Rate at Elementary Stage in India 1960-61 to 1997-98

Year	*Primary (I-V)*			*Upper Primary (VI-VIII)*		
	Boys	*Girls*	*Total*	*Boys*	*Girls*	*Total*
1960-61	61.7	70.9	64.9	75.0	85.0	78.3
1970-71	64.5	70.9	67.0	74.6	83.4	77.9
1980-81	56.2	62.5	58.7	68.0	79.4	72.7
1990-91	40.1	46.0	42.6	59.1	65.1	60.9
1997-98*	38.2	41.3	39.6	50.7	58.6	54.1

* provisional

Source: DoE, MHRD, GOI/IER-2002

Although there has been steady decrease in the dropout rate, but it still stood in 1997-98 at 40 per cent in primary stage and as high as 54 per cent in upper primary stage. The dropout rate of girls was more than the boys in 1997-98 by 3 per cent in primary stage and 8 per cent in upper primary stage.

Against this background the State Education Ministers' Conference was held in October 1998 to find out how to give thrust to elementary education across the country. The Conference recommended for starting a special scheme by name 'Sarva Shiksha Abhiyan (SSA).' The scheme actually started from November 2002. The goals of the scheme are: (i) All 6-14 age group children to be enrolled in school/Education Guarantee Scheme center/ bridge course by 2003; (ii) Completing five year primary education by every child in the age group of 6-11 by 2007; (iii) All children in the age group of 6-14 to complete eight years of schooling by 2010; (iv) Bridge all gender and social category gaps at primary stage by 2007 and at elementary education level by 2010; (v) Universal Elementary Education retention by 2010 and (vi) Focus on education of satisfactory quality.

The special features of the programme are: (i) Focus on children of weaker sections especially girls; (iii) Free textbooks for girls; (iv) Special coaching/remedial classes for girls and a congenial learning environment; Back to school camps for out of school girls; (v) Teachers' sensitization prcgramme to promote equitable learning opportunities; (vi) Special focus for innovative projects related to girls' education; and (vii) Recruitment of 50 per cent female teachers.

The expenditure on the programme will be shared between the Central Govt. and the State Govt. on 75:25 basis during Tenth Plan and 50:50 basis thereafter. The programme covers the entire country and the needs of 192 million children in 11 lakh habitations. 8.5 lakh existing primary and upper primary schools and 33 lakh existing teachers would be covered under the scheme. The programme provides for opening new schools in villages which do not have schooling facilities and augmenting facilities in the existing schools by way of providing more classrooms, drinking water and toilets etc. It also provides school improvement/maintenance grants, academic support, and additional teachers to strengthen the existing staff and training for teachers. It also envisages providing computer education. The programme seeks full participation of the community and education plans to be prepared in consultation with Panchayati Raj Institutions.

Notwithstanding best intentions and above provisions in the programme first target itself remained yet to be achieved.

3. SCHOOLS AND SCHOOL SYSTEMS

1. Education at Home

Even before child goes to school or even during schooling, learning at home makes much more difference. Children are very keen in their perception and they always perceive their parents as role models. Parents have therefore to set the example if they wish their children should pick up good habits. For example if father has a smoking habit, it is very difficult to prevent child from imitating this habit with the only difference that out of respect or fear of the father he may not smoke in his father's presence.

At home, child can be taught about cleanliness, sanitation, and tidiness. He can also be taught about ethics i.e. good behaviour, truthfulness, honesty, non-injury to others etc.

This will lay good foundation for holistic development of the child and deriving maximum advantage from the education which will be taught in the school.

Parents should express love towards their children and take proper care of them so that they are kept away from any unsafe or undesirable situations. However they should avoid pampering their children. Pampering makes the child attention-seeking and such a child will find difficulty in adjusting with other children when he goes to school.

2. Anganwadi Centres

In villages, most of the men and women have to work in the fields. In view of this, proper childcare becomes a problem for them. Further it is not practicable to send children of the small age group away. This has given birth to the concept of *Anganwadi*. Anganwadi, which is fully funded by the State, is a para-school in the neighbourhood. It is supposed to have its separate accommodation, teaching aids, *Sevikas,* meals and facility of regular health check ups. Where separate accommodation is not available, it is temporarily situated in the house of *Sevika.* There are altogether 6 lakh Anganwadi centers manned by 12 lakh *Sevikas* throughout the country.

The Anganwadi takes care of children below 6 years and prepares the foundation for the children for taking up formal education in the primary schools which admit children from the age of 6 years. Since in this age group, children exhibit more of their creativity, besides nursery education, activities like painting, singing, games etc. are encouraged. Emphasis is also on health and hygiene of the children. The functioning of the Anganwadis depends entirely on the sensitivity of the *Sevikas* towards child development and their caress with motherly warmth. As a motivation to Anganwadi *Sevikas/Workers,* a provision of social security to them in the form of insurance scheme is on the anvil.

3. Formal School Education

Primary school caters education needs of the age group 6 to 11, upper primary/middle school 12 to 14, Higher Secondary 15 to 16 and Senior Secondary School/Junior College 17 to 18 years. Taking into account large number of out of school children, the number of schools available are not enough. Table 2.6 shows the availability of primary and upper primary schools from 1990 to 1997.

Table 2.6: Number of Primary/Upper-Primary Schools 1990-97

Year	*Primary*	*Increase*	*Upper-Primary*	*Increase*
1990	560,935	57,172	151,456	29,079
1991	566,744	5,809	155,926	4,470
1992	571,248	4,504	158,498	2,572
1996	598,354	27,106	176,772	18,272
1997	610,763	12,409	185,506	8,734

Source: DOE, MHRD, GOI/IER-2002

Apart from proper buildings, these schools are supposed to have, experienced teaching staff, all basic facilities like furniture, teaching aids, drinking water, toilets, library, play grounds, sports items, regular health check up etc. Like Shantiniketan, founded by Maharishi Rabindranath Tagore, these schools should have around them an environment which is friendly towards children, an aura of being nearest to the nature, that is a place where there are small and big trees and lush greenery.

4. Non-Formal Education (NFE)

The scheme of Non-Formal Education was first started in 1979-80 to target the children in the age group of 6-14 years who had remained outside the schools having formal education system. Besides rural area, the scheme also coveres urban slums, hilly, tribal and desert areas. The assistance for implementation of the programme by the Central and State Govts. has been in the ratio of 60:40 for running co-educational centers, 90:10 for girls' centers and100 per cent assistance to Voluntary Agencies.

In order to provide the scheme as a popular alternative to formal education, it has been revised and named as Education Guarantee Scheme and Alternative and Innovative Education (EGS & AIE). The scheme envisages covering all the habitations where there are no education centers within a radius of 1 km. and at least 15-25 children of 6-14 years age group who are not going to school are available. The scheme forms part of the overall national programme of Universalisation of Elementary Education (UEE)/Sarva Shiksha Abhiyan. The pattern of sharing the cost of implementation of the scheme has been made uniform and the same is in the ratio of 75: 25 between Central and State Governments. The scheme has started operating from 1st April 2001.

5. Common School System (CSS)

Presently there are number of schools which are based upon different denominations like religion, language, caste, creed, financial status etc. With a view of removing such divisions in the formative age of the children, the concept of common school system or its nearest form of neighbourhood school was first mooted by the *Kothari Commission*. The idea is to establish schools which will be funded by the State but managed either by private body or local state body. The education which will be of good standard will be accessible to children of the whole community irrespective of any denomination such as economic status, caste, gender, disability, creed or religion. Apart from easy accessibility and spread of literacy, adoption of this system would work towards laying solid foundation for national integration. Early age of the child is the best time for inculcating good ideas as the receptivity and the retention are at their best at this formative stage. Once children come of age, it becomes difficult to change ideas already entrenched in their minds though they might be wrong. Common school system however should not be misunderstood with uniform type of education. Rather here the word common denotes a common equal quality of education. In these schools, there would be a conscious effort to enable a child to become a good citizen in future.

However, unless standard of common school system is good it is impossible to assume that children of all communities would opt for these in preference to their own community schools. Since cost of providing good education is more, this system is not picking up as the resources of the State are already tied up with conventional schools.

6. National Bal Bhavan

National Bal Bhavan under the agies of Bal Bhavan Society of India was established in 1956 with the aim of providing opportunities to children in the age group of 5-16 years to enhance their creativity whatever social strata they may belong to. The children can pursue the activities of their liking such as creative arts, performing arts, photography, physical activities, experiments in science etc. These creative activities will help in all-round development of children including scientific temper. At present there are 73 Bal Bhavans across the country.

7. Navodaya Vidyalayas

In rural area, since primary concern is to provide more outreach of education so as to spread literacy, the quality aspect is not given that much priority. Further, the talented and bright students in rural area experience difficulties in going to cities where quality education is available. In order to provide easy access to such children, the National Policy on Education-1986 envisaged setting up of model school one in each district. Named as Jawahar Navodaya Vidyalayas, these are fully residential co-educational institutions. There are now 506 schools covering as many districts spread over 34 States/ UTs with enrolment of 1.58 lakh students as on 31.3.2004. Girls' enrolment is specially encouraged so that at least 33 per cent of the students enrolled are girls.

8. Kasturba Gandhi Swatantra Vidyalava

The literacy among girls is much lower than the boys especially in rural area. Besides, on account of the fact that not every village is having a school and the poor cannot afford to send their girls away for education, the scheme namely Kasturba Gandhi Swatantra Vidyalaya is mooted. This scheme envisages building 500 to 700 residential schools for under-privileged girls.

9. Education Based on Financial Status of Parents

The schools in our country can be broadly classified in four categories. The first category is State run schools which are totally funded by the State. These include also schools run by the municipalities/municipal corporations in towns/cities and constitute largest in terms of number of schools as well as number of students. Although as per the rules teacher to student ratio in these schools should be 1: 40, in most of the cases wherein large number of students are seeking the admissions, the ratio is 1:50. Since State does not have resources to fund these schools adequately, these are not well equipped in terms of proper buildings, furniture, teaching aids etc. Most of the teachers in the schools situated in the villages do not live there. They live in towns, cities or where they have property. This affects punctuality and regular attendance of teachers and in turn the quality and content of teaching. However important feature of these schools is that there is more personal sympathetic chord between the teacher and the students.

The second category is State aided private schools. Many of these are run by politicians and social organisations. Besides imparting education, there is an element of profit motive in running these schools. However these schools play a very important role in spreading the outreach of education and thereby augmenting the efforts of the State. The teachers in these schools carry greater responsibility towards students and the discipline in students is better.

Unfortunately in both the above categories there is a lack of proper transparency in the matter of appointment of teachers. It tends to be politically oriented and in many cases money plays greater role.

The third category is public schools which are unaided. These are situated in and around towns and cities. These offer Montessori teaching in the foundation classes and State or Central Board pattern in the higher classes. These schools are better equipped for imparting the education. Provisions of computer facility and special classes for the students who are weak in studies are some of the good features. The teachers are more committed towards students' performance. These schools, besides education, lay fairly good emphasis on development of wholesome personality and discipline. The fee structure is fairly high in these schools. In spite of this, there is increasing trend in not only well to do but also in middle class families to put their wards in these schools. Interestingly lower income groups are also veering round to put their wards in these schools even if it may mean cutting down expenditure on other fronts or taking up additional assignments for sustenance of the family.

The fourth category is ultraexpensive schools that attract ultrawealthy families to put their children in these schools. The comfort level in these schools can be compared with five star hotels. The motive of putting their wards in these schools vary widely.

Some families look at these schools as status symbols. Some put them because they have so much wealth that this is perceived as better way of spending the money. Some do it because they want to seclude their wards from coming into contact with the children of poorer strata of society. Some do it with the idea that the children will be better equipped to be sent to abroad for higher studies.

In these schools the syllabus is generally Indian School Certificate Examination pattern and curriculum is oriented on the lines of European or Western countries. The school building and transport buses are fully air conditioned, every child will be provided a laptop, canteen will be like a good restaurant, the campus will have sports facilities ranging from horse riding, golf, snooker, swimming pool to conventional games like cricket, hokey, football, basketball etc. It will have amphitheatre for holding cultural programmes, health centre, gymnasium, shopping plaza etc. The expenses for each child will be like Rs. 5 lakh per year or even more.

Although facilitywise these schools are excellent, they suffer from some distinct disadvantages. Extra-richness can be associated with nostalgia of different kind. Further, in future life, they will feel uncomfortable to adjust with the Indian environment. They can however, be useful for filling the positions in their family business or living abroad.

10. Changes in Education System from Past

The process of education has undergone a sea change over a period of time. A few decades earlier, education used to be a fun for the child. Admission to any fairly good school or even collage was a routine affair. There was no semester system then. It used to be fun together with study for initial 8 months from starting of the session and only during 2 months prior to examination, one was required to study hard. There was again a period of summer vacation for enjoyment. Also one could spend considerable time everyday for playing games in areas near residence or in school grounds.

However, situation has drastically changed over a period of time. The child now has to carry to school a big bundle of books and copies enough to bend his spine. He is given so much homework that neither he nor his parents get any leisure time. There is a need to design education system in such a way that while gaining knowledge by a child, his personality also develops. Today not only students are reluctant to participate in sports but even parents do not encourage them for this, as both feel, the time spent on sports can be better utilised for studies. This way not only children do not develop their wholesome personality but also they ignore the fact that one will be able to concentrate better in studies if he spends at least some time in sport.

Presently up to 10th standard, there are no vocational courses. The students therefore tend to become theoretical. Although students at young age have creativity, they are unable to develop this faculty. Vocational course, if introduced at early stage, will broaden their horizon and increase their confidence level to undertake self-employment.

4. CURRICULUM OF STUDIES

1. Medium of Instruction in Schools

It is far better if the medium of school education is mother tongue. Since child is accustomed to the use of mother tongue in home environment, he will feel more at ease with that language in school. This many however be all right up to certain standard. Since one has to be groomed for working in future at national or international level, process of imparting knowledge of Hindi and English has to start as early as possible. This would no doubt increase burden on the child as he has to learn three languages if mother tongue is other than Hindi or English. However, the children have tremendous adaptability and in any case there is no alternative to this in existing as well as in future scenario. Hindi is understood by most of our countrymen and therefore it has become our national language

In global context, English is commonly used for all purposes. In past, efforts were made by eminent linguists in some of the countries especially France in developing a language as an alternative to English. This was attempted when there was no globalisation process in existence. The move however did not make much headway. Presently globalisation picking up fast, it is impossible to create an alternative. There is also a fast development in technology and scientific research and much of the literature in these fields is in English language. With World Trade Organization in place, trade barriers between the countries are fast breaking down. In this scenario, all the countries are almost reconciled with accepting the English as a sole medium of international communication practically for all purposes. We are witnessing a change in the outlook of our countrymen also in favour of using it extensively. Earlier inhibition of not using it because it is a foreign language has more or less disappeared. It is a well recognised fact that because of good knowledge of this language, Indians have better acceptability over others i.e. Chinese, Japanese etc. for jobs, higher studies, business opportunities etc. in USA, Europe or other developed countries. The large population of Indians in USA, UK etc. is proof of this acceptance. In the context of reaping high benefits in world arena due to good knowledge of English, we should move forward in acquiring further excellence in this language. Although we may refuse to acknowledge openly, it would not be wrong to say that this is one positive gift of the English rule in India.

2. Curriculum

Curriculum of Education has to be so designed as to achieve the core objectives of the education. Apart from the knowledge of the letters, it has to cater to basic factors such as character building, encouraging creativity etc. Emphasis has therefore to be on imparting knowledge through work experience rather than books alone.

We have also to keep in mind changing scenario around us. As time is passing, due to fast communication, the world is becoming smaller and process of globalisation is taking place. This means we cannot remain isolated any more and there is a necessity to learn interpersonal relationship during formative period of life. The system has therefore to be oriented in such a way that there is an assimilation of the minds of the children coming from different socio-cultural-economic background.

3. Environmental Studies in School Education

Many problems are cropping up out of unawareness about hazards that can occur in nature. The National Policy on Education (NPE)-1986 provides that the protection of environment is a value which, along with certain other values, must form an integral part of curriculum at all stages of education. In order to achieve the desired results, there is a need to start environmental education right from the earliest stage so that the children are sensitised to the hazards inherent in nature upsetting the ecological balance in nature. This step includes awareness and respect among students for basic concepts relating to the conservation of environment.

With the above objective, a Centrally sponsored scheme, "Environmental Orientation to School Education" was started in 1988-89. The scheme envisages assistance to voluntary agencies for conduct of experimental and innovative programmes aimed at promoting integration of educational programmes in schools with local environmental conditions. Apex court has now ruled that environmental studies should be made compulsory in the school education system.

4. Science Education in Schools

In line with the National Policy on Education-1986, a Centrally sponsored scheme, "Improvement of Science Education in Schools" was launched in 1987-88. Under the scheme, states are assisted for provision of science kits to upper primary schools and assistance is also provided to voluntary agencies for conducting experimental and innovative programmes.

One of the important components of this scheme is participation of Indian students at school level in the International Science Olympiads.

5. Population Education in Schools

The National Population Education Project was launched in April 1980 with a view to institutionalise population education in the school education system. This is an externally aided project which is fully funded by United Nations Population Fund. This project is also being implemented in Higher and Adult Education Sector. The Fourth cycle of the project was completed in December 2002. From 2003, the project is to be funded by the Central Govt.

6. Religious Teachings in Schools

There is ongoing debate in our country as to whether religious teaching should form part of the curriculum. Religion is a matter of personal faith and therefore there is bound to be a difference of opinion in this matter. However there is a broad consensus among the people about introducing ethical values which are common to all religions, which include truthfulness, honesty, cleanliness, purity of body and mind, good conduct, helping others in need etc.

7. Burden of Studies

Gone are the olden days when children used to enjoy everything that was around: laughter in the family, playing in the fields, rain, beauty of rainbow and so on. These were bringing enormous joys to children. No pressure of studies used to be felt. There was no weekly or monthly test, only examination at the year end for which last two months' study was enough. The process of education was free from any feeling of compulsiveness. Today grinding mill of the studies starts right from Nursery classes. Weekly or monthly tests in primary classes onwards have become common. Today children are required to carry everyday to school, a burden in the form of books and copies weighing up to 5 kgs on their back. No wonder sooner or later they develop back pain due to this load. The extent of homework given is so much that it leaves no leisure for other creative interests like playing, reading a storybook, singing a song, drawing a picture or watching a bird. Entire day's schedule is so tight that children find practically no flexibility. This is nothing but regimentation from the early age. They often carry a feeling of inadequacy that something told by the teacher is not done as would meet the expectation of the teacher i.e. a feeling of inadequacy.

Unfortunately parents also harbour unrealistic expectations from their children. Those children who do not get high grades in the examinations are despised both by teacher as well as parents. The children look for somebody who can listen to their side of story, understand their feelings. But in today's nuclear families, there is no *Dadi* or *Dadaji* who can provide not only love to the child but also on occasions prevail upon child's parents to be realistic about the expectations from them. Children thus feel psychologically themselves as enclosed in an envelope. In many cases, situation worsens so much that it leads to tragedies. It is improper to put blame on the innocent child for such happening. The responsibility lies with the parents, teachers and the education system which drive him to adopt the final route of escape. Leave extreme cases, generally the existing system rather than helping towards personality development, actually hinders the process.

Regarding home work, ideally there should be no need to carry much of load of books and copies everyday to school and back. Most of the studies should be completed in the school itself. This will provide enough time for the children to pursue hobbies of their interests. In fact some of the schools, catering to elite class, have already started adopting this practice. In some schools there are two sets of books, one to be left in the school.

There is a way to make the learning process interesting. Teacher has first to create friendly atmosphere in the class. He has also to know about the children's psychology of receptivity. He should accordingly orient his method of teaching in such a way that the children get interested in the subject.

A practical example here can substantiate the above point. Mathematics is normally taken as the toughest subject and students are fearful about it. In 1955-56 during our class X and XI, there was one Math teacher, quite strict otherwise, used to start the class first by narrating some interesting story. All students used to get engrossed in his story. After finishing the story, quietly he used to change over to Math subject. He used to involve all the students in the process of finding solution to the problem. He will project the whole exercise in such a way that the problem was solved by the students themselves as if without any aid from him. He never tried to take credit for himself for solving the problem. Thus students used to carry a feeling of achievement. Attendance in his class used to be full and results were also excellent. The crux is that instead of starting with a sermon, teacher has to first create interest in the class to accept what he wants to teach.

Some teachers are very harsh in their approach. Students develop negative feelings towards such teachers and they try to avoid their classes. Even if they attend, they do it as a matter of compulsion. Similar friendly approach if adopted by the parents, they will find their children doing extremely well.

8. Scourge of Tuitions

The concept of tuitions originated from the idea that some of the students for various reasons are so weak in studies that without extra efforts from the teacher they would not be able to cope up with the class. Ideally the responsibility of help to weak students should either be taken by their own schools by way of arranging special classes or by the parents if they are educated. However over a period of time it is seen that teachers with the motive of earning extra money started giving tuitions in their homes. Gradually the deterioration went up to such an extent that teachers prefer to avoid teaching properly in the classes and instead coaxing the students to take to private tuitions.

These days there is a mushroom growth of coaching classes some of which dupe the students by way of giving out attractive advertisements with false claims about the results. In this competitive world, students are a worried lot and many of them fall prey to such institutions with the hope that they would be better off with the tuitions. Since private tutors as well as coaching institutions work purely on commercial lines, they are not really worth the time, effort and money. Those teachers who neglect their basic responsibility of giving proper teaching in the classes are doing great disservice to the cause of education. Further, it is a responsibility of the school management to arrange special classes in the school itself to help the weak students. Parents who are capable of teaching have also a responsibility to help their children instead of putting the whole burden on the schools.

9. Study/Work Groups

It is an interesting method to divide the class into few study/work groups and giving them a simple problem to solve. The competition between the groups will highlight the importance of working as a team within the group. The students will thus understand basics of the interpersonal relationship which will prepare them for better interaction with others in the society.

5. DIFFERENT APPROACHES/EXPERIMENTS

The task of providing Education for All in our country is so gigantic that no one particular approach will bring desired result. This requires multilateral and multilevel approach. Many organizations have come up with an intention of contribution to the cause of helping to spread literacy in rural area. They have attempted to involve community itself in their effort. The work of some of these organizations, which is quite notable, is narrated here.

1. Lok Jumbish

Lok Jumbish is an innovative scheme started in Rajasthan in 1992 with the assistance of Swedish International Development Agency (SIDA) with the objective of providing education for all through community's participation. The sharing of funds by SIDA, Central Govt and Rajasthan Govt. was in the in the ratio of 3:2:1. The third phase of the scheme to be completed in June 2004 is financially assisted by the Department for International Development (DFID), UK. Project innovations were undertaken in several aspects of primary education like teachers' training, minimum level of learning, opening of new schools, upgradation of existing primary schools, opening non-formal education centers, school mapping etc. The first phase of the project was implemented during June 1994. The second phase was implemented between July 1994 to December 1999 and the third phase was from July 1999 to June 2004. The programme covered 13 districts, 177 blocks and 23,234 villages.

The Fifth Joint Review Mission (JRM) visited Rajasthan in January 2004, to undertake a review and appreciated the progress of the project. The JRM recommended that the good processes and practices of LJP be mainstreamed after integration of the project in Sarva Shiksha Abhiyan.

2. Mahila Samakhya

The role of rural women in the development process in rural area is well recognised. Mahila Samakhya (Education for Women's Equality)

Programme started in 1989. provides for the education and empowerment of women in rural area, particularly women from socially and economically marginalized groups. It is being implemented in 13,000 villages in 60 districts of 9 states.

The main objectives of the programme are: to enhance the self-image and self confidence of women; to create an environment where women can seek knowledge and information which empower them to play a positive role in society; to establish a decentralised and participative mode of management; to enable the group to actively assess and monitor educational activities in the villages; to provide opportunities for education for women and adolescent girls and to bring about greater participation of women and girls in both formal and non-formal education schemes.

The organisation provides a forum where women can meet and discuss their problems. A group of two or more women called *Sakhi* or *Sahayaki* are trained to work as catalysts. The *Sahayoginis* who look after a group of 10 villages work as motivators, supporters and guides.

3. Janshala

The progress of literacy has not been uniform in all the states. Likewise there are differences in the rural area of the state itself. Janshala (GOI-UN) is a collaborative effort of the Govt. of India and UN agencies to provide support to the ongoing efforts towards achieving UEE. It seeks to involve local community in primary education programme and is aimed at making primary education more accessible to girls and children in deprived communities, marginalised groups, SCs/STs/minorities, working children and children with specific needs. It is a block-based programme and the blocks have been selected on the basis of indicators such as low female literacy, incidence of child labour, concentration of SC/ST population etc.

UNDP, UNICEF and UNFPA have committed funds for the programme and UNESCO and ILO have offered technical know-how.

The programme is in operation in 139 blocks in 30 districts in nine states. Alternative schools have been set up and number of teachers have been provided training on different aspects of teaching.

4. Kerala's E-literacy Model

Kerala's Malappuram district has become the role model in the state's e-literacy drive. Project Akshaya, the district's e-literacy model has greatly upped literacy levels in the district. The district which previously recorded highest school dropouts among girls, now has taken to hundreds of Akshay centers imparting e-education, distance learning and many other services. The module covers a 15 hours training session in 40 centres located in the district. The delegations of other countries who visited the district have been so much impressed by the project that they expressed plans to replicate it in the rural areas of their countries. The success of the project has shown that the potential of rural population is no less than the urban population and it is only a question of providing opportunities to rural people for expressing their creativity.

6. SPECIAL PROVISIONS FOR PROMOTING EDUCATION TO WEAKER SECTIONS OF SOCIETY

Large disparities in any field between various sections of society is the main cause for disharmony. Since education is a great leveler and plays pivotal role in the progress of a person, special provisions have been made in the Constitution as well as policy framework to provide greater opportunities of education to weaker sections of the society.

1. Special Provision for Education of SCs/STs

Govt. has taken number of steps in promoting education to SCs/STs. However there is still a long way to go for these communities to attain the educational level of other communities. In line with the National Policy on Education-1986 and the Programme of Action-1992, the following provisions have been incorporated in the existing schemes of the Elementary Education and Literacy: Relaxed norms for opening of primary/upper primary schools. A primary school within 1 km walking distance from the habitation of population of 200 instead of earlier norm of 300; Abolition of tuition fee in Govt. schools at least up to upper primary level; Incentives like free textbooks, uniforms, stationary, schoolbags etc. The Mid-Day Meal scheme is viewed as an incentive to the success of the programme. It covers all students of primary classes in all the Govt. and local body and Govt. aided schools with a view to improve enrolment, attendance and retention. The scheme is simultaneously aimed at improvement in the nutritional status of the children.

The scheme viz. Sarva Shiksha Abhiyan (SSA), District Primary Education Programme (DPEP) and Shiksha Karmi Project (SKP) lay emphasis on providing infrastructure facilities and special interventions for the education of girls and SCs/STs.

2. Education for Minorities

The National Policy on Education-1986 and Programme of Action-1992, with a focus on equity and social justice, envisage paying greater attention to the education of the educationally backward minorities. In line with this provision, two schemes were launched in 1993-94, which have since been merged and named as Area Intensive and Madarsa Modernisation Programme. The scheme is funded by the Central Govt. and is aimed at giving thrust to education for minorities.

3. Female Literacy: Special Interventions

The main problem continues to be how to increase the outreach of elementary education and enroll the out of school children especially girls. As per 2001.Census female literacy is 53.67 per cent whereas male literacy is 75.26 per cent. In order to target the least female literacy areas, 45 districts in UP, Bihar and Orissa were identified where the female literacy is below 30 per cent. Accelerated programmes are taken up for improving female literacy in these districts and for achieving the results women volunteers and NGOs are also involved.

The schemes like SSA, DPEP, SKP etc. are focusing on improving female-literacy throughout the country.

In UP, 97 NGOs and in Orissa, 117 NGOs are engaged in helping the sizable number of females in literacy. The interventinos cover females in UP (27 lakh), Bihar (20 lakh), Jharkhand (5 lakh) and Orissa (9.1 lakh).

4. Hostels for Girl Students

Under the new name of the scheme "Access and Equity," financial assistance is given to voluntary organisations to improve the enrolment of adolescent girls in rural area and weaker sections. Parents are reluctant to send their daughters to far away schools. To get over this problem, preference in providing assistance is given to hostels located in educationally backward districts, particularly those predominantly inhabited by SCs/STs and educationally backward minorities.

7. EDUCATIONAL RESEARCH AND TRAINING OF TEACHERS

1. National Council of Educational Research and Training (NCERT)

The National Council of Educational Research and Training was founded in September 1961 with a view to assist and advice Central and State Governments on matters related to school education. It provides academic and technical support for qualitative improvement of school education. The council aids, promotes and coordinates as also conducts research in school and teacher education, organises pro-service and in-service training of teachers, extension services for institutions; develops and experiments with improved educational techniques, practices and innovations; assists the states in developing and implementing programmes for qualitative improvement of school education; collaborates with international organisations in the field of educational research and training.

2. Education and Training for Teachers

This is very important aspect of the education system. Since children are in the initial learning stage, it is very essential to pass the teachers or para-teachers through sensitivity training before inducting them for actual teaching.

As a follow up of National Policy of Education (NPE) and Programme of Action (POA)-1986, a scheme of Restructuring and Reorganisation of Teacher Education was started in 1987-88. This is a centrally sponsored scheme aimed at creating institutional framework, academic and technical resource base for orientation, training and continuous upgradation of knowledge; competence and pedagogical skills of school teachers, adult and non-formal functionaries and teacher educators. Terming district as a unit, the scheme envisaged setting up of District Institute of Education and Training (DIET) in each district. It provides academic and resource support to elementary education teachers and Non-Formal Education (NFE)/Adult Education (AE) instructors. More than 2.3 lakh teachers have been given orientation training under Special Orientation Programme for School Teachers (SOPT) in the use of Operation Blackboard (OB), Teaching Learning Equipment (TLE) and implementation of Minimum Level of Learning (MLL) strategy with focus on teaching of language, mathematics and environmental studies. Since many aspects of education are state specific, the need for revising the scheme of teacher education so as to make it more flexible and state specific has been well recognised.

At the time of induction of new teacher or para-teacher, at least short term training should be organised for them at taluk/block or district level and if this is not possible, training can be given by the experienced senior teacher from the same school who can act as a *mentor*. This prior knowledge would prepare them to teach the children more effectively.

3. National Council for Teacher Education (NCTE)

The National Council for Teacher Education has been established by an act of Parliament in August 1995. The objective of NCTE is to achieve planned and coordinated development of teacher education system and regulation and proper maintenance of norms and standards of teacher education. As on 31st March 2004, 2995 teacher training institutions offering 3329 courses have been recognised by NCTE with an intake of 2.56 lakh teacher trainees.

8. ABOUT TEACHERS

1. About Teaching Staff

The performance of the teacher is more of a personal trait. There are both sincere as well as insincere teachers. In olden times education used to be treated as a noble profession and teacher used to be respected by student as one's parents. Teacher also used to care for the welfare of the student. However over a period of time, teaching has become for some a commercial business. Students also do not pay respect to teacher as he/she rightly deserves. There is a loss of value system on the side of teacher as well as students. Till few years back teachers were appointed only on regular basis. The security of regular service brought in its wake inefficiency in large number of teachers. Most of the teachers live in towns or at a place where their property is situated. This affected their regular attendance and punctuality in the schools, save exceptions. Teacher absenteeism has been identified as a major obstacle in achieving the goal of Universalisation of Elementary Education (UEE). Table 2.7 shows absentee rates (2003) of primary school teachers in some of the states.

Table 2.7: Absentee Rates of School Teachers (2003) in Some of the Major States

State	*Absentee Rate per cent*
Andhra Pradesh	31
Bihar	26
Gujarat	21
Karnataka	23
Orissa	14
Rajasthan	23
Tamil Nadu	17
Uttar Pradesh	26
West Bengal	21

Source: India Sustaining Reform-2003.

In earlier time, starting salary of a teacher used to be about Rs. 4000-5000/- pm. As a consequence of the implementation of Fifth Pay Commission Recommendations, the starting salary including pay and DA for regular teacher is about Rs. 7000-8000/- pm. The expenditure on salary of the regular teachers thus increased tremendously. Thus while State expenditure increased, there has been no commensurate improvement in the standard of education. However it is a matter of fact that there are quite a number of sincere teachers in regular category and their quality of teaching and experience are of very high standard.

2. Shiksha Karmis

In Rajasthan, Madhya Pradesh and some other states, under Education Guarantee Scheme there has been effort to involve parents through *shala shiksha samities* to manage the education centers and employing

Shiksha Karmis or para-teachers on performance contract basis. With great emphasis on various programmes, aimed at achieving 100 per cent literacy at the earliest, and against, the backdrop of limited resources available with the State, and to address the problem of absentee teachers, it is generally found expedient to appoint para-teachers or *Shiksha karmis* on contract basis instead of regular teachers. The expenditure on the salaries of the teachers, which is a major component in the school education, is much less in this system. Hence the appointment of regular teachers is almost stopped. The para-teacher is paid Rs. 3000/-pm and there is no payment during the vacation. In another system a para-teacher is called for taking specific periods and paid Rs. 60/- to 70/- per period.

The appointment of para-teachers has been delegated to village level bodies who normally prefer local persons. This has improved availability of the teachers in the schools. Further, participation of the village level bodies in school affairs is being encouraged. The functioning of the schools is thus less affected by the centralised system of bureaucracy. The introduction of *Shiksha Karmis* in the education system demonstrates that even when they are working in difficult circumstances, their motivation can be sustained by effective training, sensitive nurturing, community's co-operation, regular participatory review and problem solving approach. The advantages of *Shiksha Karmi* system has attracted attention of all the states and many of them have already started implementing the same.

There is evidence of improvement in the attendance of the teachers as well as results at the lower level classes. The success of this system however depends upon the quality of the teachers locally available and how far village level bodies avoid temptation of foisting persons of their choice, disregarding claim of more qualified persons available. In this context it is very necessary to monitor performance of this system regularly from the point of view of long-term gain both in terms of enhancement in literacy and quality of education.

9. SPECIAL ROLES

1. Role of Non Government Organisations

The task of achieving 100 per cent literacy which is the goal before the country is so gigantic that it requires multifaceted approach. Surely it cannot be achieved by the efforts of the Govt. alone and the contribution of all available institutions organisations will have to be mobilised. It is in this context that the role of NGOs becomes important. The National Literacy Mission fully recognizes the potential of the NGOs in furthering its objectives and has taken measures to strengthen its partnership with them. It has assigned them an active promotional role in the literacy programme. Besides imparting literacy, NGOs also provide academic and technical support through research and innovative programmes; conducting evaluation studies; organising seminars etc. Throughout the country there are 26 State Resource Centres (SRC) managed by NGOs, who provide these services.

2. Role of Information Technology (IT)

IT can play a major role in the field of education. For increasing the rate of literacy, besides shortage of classrooms, textbooks and other facilities, major factor is shortage of qualified teachers. IT has a potential to bridge the gap significantly. The distance e-learning modules can do the job of teaching to a great extent. In fact the children may find some of the interactive modules more interesting. The modules have also distinct advantage of easy updating as against updating the knowledge manually.

10. ORIENTATION OF EDUCATION TOWARDS SKILL DEVELOPMENT

1. Place of Sports in Education

Sport contributes not only to the physical development but also mental development. It also teaches importance of teamwork. It is beneficial for the student to spend at least 1 to 2 hours daily in sport's activity. It gives an opportunity for oneself to assess his physical as well as mental capability and for improving them further. It also creates a feeling of courage and confidence. It is a mistaken belief that participation in sport will detract attention of the student from the studies. In fact playing games for an hour or two will refresh him and enhance his ability of doing studies better.

2. Uneven Focus on Subjects

In the existing system, the focus is on teaching of theoretical knowledge that too emphasis is on certain subjects such as physics, chemistry, math and biology. Such emphasis stems from the general expectation of the parents to send their children to professional colleges for which present system of entrance test requires extensive theoretical studies. In this process, subjects like humanities, social sciences etc. are not given their due importance.

3. Education Towards Skill Development

There is a need to orient our education system in such a way that in the system both theory and application part are properly balanced. A beginning has been made by way of introducing computer subject as well as its use in public schools. Facility on these lines, if not similar, is to be provided in State run and State aided schools.

But let us not be enamored too much by the limited knowledge of the use of computers. Many use them for playing video games. It would do good to them if instead of playing video games, they are encouraged to play in the open field. By this, at least their respiratory system and stomach will become stronger.

The application part of the education can include experiments in the laboratories, learning how simple repair jobs can be done, working on crafts, practicing use of safety appliances etc. It can also include physical demonstration of cooking food. After all, food is one of the most important part of our livelihood. By doing these activities with their own hands they will develop not only their skill but also confidence in themselves.. They will thus be better equipped for solving the problems that they will face in life when they are grown up.

Because of large population and limited job opportunities, there will always be a difficulty in getting a secure job in the organized sector. This difficulty can be partly overcome by encouraging students to take up vocational training. In the present circumstances quite a number of students may opt for taking up formal courses in Industrial Training Institutes and then go in for self employment. In course of time they can themselves become service providers.

Application part of the education process will thus act as a catalyst to produce students with practical knowledge. There is a proposal to introduce a separate stream of vocational. education, parallel to academic stream, from class IX onwards in place of class XI as per the present policy. Pros and cons of such a proposal has to be studied in depth. The proposal has an advantage of attaining vocational skills at an early stage but creating separate stream below matriculation may prevent students who may inadvertantly take up this stream from reverting to the normal educational channel.

11. CASE STUDY OF EDUCATION FACILITY IN A VILLAGE-NATALA

In the village Natala, there are 150 houses with a population of 625 people. Earlier there was a Balwadi run by one *Sevika.* The Balwadi is now upgraded to an Anganwadi run by two *Sevikas*. It looks after 30 children and gives nursery education to the children. There is also a Zilla Parishad (ZP) school in" the village up to Middle School i.e. VII standard. Mahadeo, a student of class X has to cycle 5 km to attend the higher secondary school. Due to non-availability of regular public transport, girls are compelled to stop their studies after middle school. Even for boys like Mahadeo, their studies are interrupted during rainy season, as they are unable to cross the *nallah* passing nearby the village as it overflows with water.

There are altogether 100 students out of which 70 are studying in primary classes and the rest in the middle school. There are 5 classrooms as against 7 classes. The children of two classes have to be accommodated in the same rooms in which other classes are running. The number of teachers (3) is just half the sanctioned strength (6). Whenever any teacher leaves the school, new teacher does not join easily.

The teachers posted in the school do not reside in the village. One of them lives in the nearby village and the rest in the nearby town. Therefore they leave the school according to the availability of public transport. Many of the children drop out at IV or V standard. Some years ago the same thing has happened to their elder brothers and sisters who then started helping their families to eke out a living.

The Mid-day Meal Scheme has however brought joy to the children and created magic in improvement in their attendance in the school. *Usal (mixed Dal)* is provided in the Aanganwadi and *Khichadi* is provided to the primary class students.

Some children although weak in studies are just promoted to the next class. Such children get stuck later at the Board level examination. The parents feel their children becoming a doctor or engineer is a distant dream.

The illiteracy in general is the main stumbling block for

improvements in the village. The new generation is however more aware about the importance of the education. Villagers hope, once their children grow up, there will be a change in the development aspect of the village.

12. STRATEGIES TO BE ADOPTED TO ACHIEVE THE GOAL OF EDUCATION FOR ALL

1. Effectiveness of Policies and Programmes

With a view to provide Education for All, under National Policy on Education several programmes were launched from time to time. During more than 57 years of Independence, various programmes did achieve certain degree of progress in the field of basic education such as:

1. At the time of Independence, overall literacy rate was 18 per cent which has now risen to 64.84 per cent. Similarly gross enrolment ratio which was 43 per cent is now about 80 per cent.
2. The rate of growth is more in rural area than in urban area. This is due to the efforts of the Govt. and Non-Govt. organisations in improving the literacy as well as because of the fact that the literacy rate in rural area earlier was too low.
3. The gap in male-female literacy rate has decreased from 24.84 per cent to 21.7 per cent. This again is due to the efforts of the Govt. and Non-Govt. organisations in improving the literacy in females as well as because of the fact that the literacy rate of females earlier was too low.

Though there are 118.4 million children enrolled in 2002-03 in elementary education [72.7 million in primary level (class I-V) and 45.7 million in upper primary level (class VI-VIII)], there are over 100 million children who are still out-of- school.

It is therefore clear that while the achievements due to the efforts of the Govt. and Non-Govt. organisations in improving the literacy are not small, the fact remains that the goal of achieving Education for All is still far away. It would be impossible to achieve the goal of Education for All, which is by no means easy, without adopting necessary strategies. These have been mentioned at the places where various educational Programmes/Schemes have been discussed. However main strategies are discussed below:

2. Enrolling out-of-School Children

According to 1991 Census, there are 11.28 million child workers (6.18 million boys and 5.10 million girls) in the age group of 5-14 years. About 91 per cent of these children are concentrated in rural area. Further as per Govt's own statistics, there are about 100 million out-of-school children in the age group of 6-14 years by the end of 2003. Enrolment of these out-of-school children is dependent on various factors such as parental willingness and encouragement, their perception of the value of education and their capacity; availability of schools, availability of infrastructure therein, regular functioning of the schools, etc. Needless to say poverty has been the major stumbling block in getting into school those out-of-school children. Parents in poor families feel their children especially boys can help in their fieldwork and girls can help the family chores at home. For very poor families, even arranging food is a problem. In this context, the introduction of Mid-Day Meal Scheme has done wonders. Besides economic compulsions, there are social constraints in getting children from SCs and tribal areas enrolled. The aforesaid factors are detailed over hereafter.

3. Providing Adequate Number of Schools to Improve Accessibility

As per the distance norm provided in the Education Policy, all habitations with a population of 300 each should have primary school within a distance of 1 km and upper primary school within 3 km. The norm of population of 300 has since been changed to 200 While India has 11 lakh habitations, the number of primary and upper primary schools are 8.5 lakhs. Although there has been considerable progress in establishing new primary schools, there are still 16.64 per cent of habitations which are yet to be provided the facility of primary and 24 per cent of habitations with upper-primary education. The shortage of upper primary schools is more acute. The changing ratio of upper-primary to primary schools from 1: 6 in 1957 to 1:3 now indicates definite improvement. However this ratio has to be brought to the level of 1:2.

It is thus evident that there is a need to open new primary as well as upper-primary schools even to fulfill the existing distance norm. It is ironical that while there is a need to open new schools, there are instances of closing down of some of the newly established schools. This is happening normally due to non-availability of teacher or minimum infrastructure required.

While emphasis on primary and upper primary education has to continue, there has to be corresponding increase in opening of schools and creating infrastructure for classes IX to X as otherwise it will be a desperate situation in 2007 when half of the 3 crore students will have no school to go after class VIII.

4. Providing Adequate Infrastructure

The story of paucity of infrastructure starts right from the stage of Anganwadis. A survey conducted by the National Council of Applied Economic Research (NCAER) of infrastructure facilities shows more than 40 per cent run neither from buildings of Integrated Child Development Scheme (ICDS) nor rented premises but often from the homes of Anganwadi workers. Most lack even basic facilities like drinking water, toilets etc.

Some schools have thatched roofs which are vulnerable to fires. In one of such schools in Chennai, more than 90 school children died on 18th July 2004 due to fire which started from the adjoining make-shift kitchen. In December 1995, about 400 children died in a fire in Dabwali, Haryana. Many of the schools are having only one classroom and therefore such schools are run on makeshift basis. From safety and learning point of view, proper building is a prerequisite for running the school.

In number of schools there are no teaching kits, leave alone sports facility. Each primary school should have at least two classrooms and facilities as mentioned. There is also a need to introduce computers in upper primary classes to lay the foundation for preparing the students for the futuristic scenario. Normally the schools in villages are run by the Gram Panchayats and they have invariably funds problem as they are not generating enough revenue by themselves. Thus whatever funds are received from the State Government, these can only be spent on the school. What is required is allocation of adequate funds for providing essential facilities in the schools. Lack of initiative on the part of Panchayat is another reason for non-availability of infrastructure.

5. Availability of Teachers and Their Performance

Since there is a severe problem of unemployment in the country, finding persons for teaching is not at all difficult. The main problem is that the teachers are reluctant to serve in the rural area. Therefore generally teachers, once posted in rural area, try for a transfer to urban area unless they have their own establishment at a place of posting. Thus his energy is fully concentrated in approaching either someone in bureaucracy or politics for a transfer to a place of his choice. This reflects on his performance as a teacher as well as non-punctuality in the school. However to be fair, this phenomenon does not pertain only to teachers but the entire service staff posted in rural area.

The situation has become so piquant that most of the States have ceased to employ fresh regular teachers and instead *Shiksha Karmis/* para-teachers are appointed to fill up new vacancies. These teachers are chosen from the locals and they are appointed on performance contract basis. The association of Gram Panchayat/Gram Sabha with important school affairs is proving to be quite useful. Another advantage of appointing para-teachers is that the salary of a para-teacher is less than one-fifth of the salary to be paid to regular teacher since the latter's salary is based upon 5th Pay Commission's recommendations. However the revised system suffers from the disadvantage that if office bearers of the Gram Panchayat/Gram Sabha act in a partisan manner in selection of a teacher or monitoring his performance, the education in the school suffers. The village community should therefore be so vigilant as to prevent such occurrences.

The performance of the teachers is likely to improve if higher authorities do not entertain undue requests of teachers for transfers and postings. Also office bearers of the Gram Panchayat/Gram Sabha should not be partisan in appointment of para-teachers and monitoring their performance. Further, participation of local community in important school matters is expected to improve the performance of the schools considerably.

6. Review of Curriculum

We are continuing education system which is almost 200 year old. It is designed to promote memorising texts and reproducing the same. A thorough review of the curriculum is therefore called for. The curriculum should be so designed that it helps to enhance not only the knowledge of the child but also his thinking power and creativity. It should also enable him to apply his mind to face new challenges in life. A beginning has been made to incorporate these elements in the system but it is still in an infant stage.

7. Quality of Education

Though govt. schools are meant to work for around 200 days in a year, they invariably function for 150 days or less. Schools are closed on various counts such as local festivals, official and unofficial reasons. The prescribed duration of the school day is 5 to 6 hours. However most elementary schools in rural area routinely start late and end early. Shorter academic years, taken together with shorter school days effectively reduce the prescribed hours of instruction almost by half. Shorter academic years and school hours, absentee teachers and poor quality teaching has a disastrous effect on the education of the poor children. The children are however promoted from class to class not on the basis of any progress or satisfactory performance but just to ensure that for less number of students in the higher class, the post of the teacher does not get surrendered. Many complete class V but are virtually illiterate. Even after more than 57 years of Independence, a literacy level of 64.84 per cent has been achieved. In rural area it is just 60 per cent. In order to spread the outreach of elementary education quickly, there is a maximum stress on enrolling as many children as possible in the schools. There is a shortage of schools/classrooms as well as teachers. After revision of salaries upward, states are finding difficulty in appointing so many teachers in regular category. They have therefore started appointing *Shiksha Karmis*/para-teachers on contract basis instead, of regular teachers. However, simple solutions have their own shortcomings. All para-teachers recruited locally are not so qualified. Further in newly opened schools, there is a paucity of necessary infrastructure. Thus while there is an improvement in literacy rate in quantitative terms, the same thing is not true about the quality. Therefore if corrective measures are not taken, the disparity between the standard of education between these children and those who attend public schools will be abysmal. In this context, the aspect of training of teachers assumes great importance.

8. Training of Teachers

There is already institutional structure in place for training of teachers in the form of District Institutes of Education and Training (DIET), Programme of Mass Orientation of School Teachers (PMOST), National Council for Teacher Education (NCTE), Block Resource Centres and Cluster Resource Centres (BRC & CRC) under District Primary Education Programme (DPEP). However main problem is the content and the processes of the teacher training. Although the number of teachers trained is high, the problems about teachers are non-observance of standards and norms with respect to entry-level qualifications, certification requirement for recruitment of teachers, duration of training, institutional infrastructure and profile of training staff.

Another development is moving away from centre based training to Special Orientation for Primary Teachers (SOPT) through teleconferencing i.e. one way-video and two-way audio mode. Distance learning as of now has come a long way. However except for initial trial in Karnataka and Madhya Pradesh, there are no serious attempts to apply the technology in training the teachers in the skill of using the technology.

The greatest challenge in the training process is sensitising the teachers about the holistic approach to child's development including his/her mind-health. Traditionally teaching is perhaps the noblest profession. There are enough teachers in our country who are sincere and want to do the honest job. If the teachers are properly trained and made accountable for the results, they can bring vast improvement in both quantitative as well as qualitative terms.

9. Removing Social Barriers Inhibiting Enrolment

Many in rural area particularly in hilly and tribal area are unwilling to send their children especially girls to school. One of the main reasons for this is a social and cultural barrier. They just like to carry on their living in old traditional way. If the schools are opened in their habitations and they are made aware of the advantages of education, enrolment is bound to improve. Similarly abject poverty is another reason. They would like to engage their children in helping the family in looking after the younger children at home or in agricultural work, as their first concern is to earn for their livelihood. In this context measures like providing mid-day meal or distribution of ration to a student will encourage parents to send their children to school. Besides, they have to be made aware about the long-term gain if their children are educated.

10. School Mapping for Planning Educational Facilities

School Mapping is an exercise to draw on the map symbolically every household and common facilities available in the village. This is a participatory exercise conducted by the members of the community and the representatives of the concerned NGO. This technique has been developed first by the NGO, Lok Jumbish, in Rajasthan. The educational status or every household member in the age group of 5-14 is indicated by small symbols. The members of the community who are a inspired group are called *prerak dal.* After discussion with the community, from the visual school mapping, the *prerak dal* prepares proposals for additional schooling facilities required and the same are submitted to the concerned authorities This system is more realistic way of planning than top-down approach and is being implemented in many parts of the country. Since the results of this system are quite encouraging, the same can be implemented in all parts of the country.

Gradually a trend is developing in villages in forming village education committee and parent teacher association. They help in monitoring of not only utilisation of the existing schooling facilities but also educational facilities additionally required. This has resulted in a shift from earlier supply based to demand based strategy for provision of educational facilities.

11. Addressing Issue of Drop Outs

In 2001-02 the dropout rate from Class I to Class V is 39 per cent. Almost half of the children who enroll in schools drop out before class VIII level. There is however improvement in the dropout rate from Class I to Class V by 3.6 per cent in 2001-02 over that in 1991-92. A positive feature is that in recent years, a reduction in the dropout rate has been faster in the case of girls than in the case of boys. The dropout rate has strong relationship with the efficiency with which the school system functions. There will be less dropout rate in the case of: delivery of good quality of education, regularity and punctuality of teachers and pupils, appointment of more lady teachers to retain girls, encouragement from teachers for continuation of education by the children. Importantly, on psychological front, if the learning in the school gives a child a feeling of excitement and self-esteem, he will be more attracted to continue the education further.

Then there is also a factor of willingness and capacity of the parents for enabling children to continue their education. At times the children drop out on account of poor economic conditions of the families, so as to help their parents for family chores at home and doing fieldwork.

12. District Level Planning

Earlier the planning for educational needs at the district level used to be done at the State level. Since local needs vary from district to district, decentralization of the planning function is the only way to make the plan more realistic. The recommendations given by the village on the basis of school mapping can form the basis for firming up district level plan, which can then be taken up for implementation. This function of planning at the district level can be undertaken by the District Primary Education Programme or by the District Education Office.

13. Mid-Day Meal Programme

This programme was started with a view that the educational development of a child cannot take place without good health, which again is dependent on nutritious food. It was concluded that the meal if served in school will go a long way to provide much needed nutrition to children coming from poor families and this would also in turn encourage children to attend the school. Since poverty is rampant in rural area, Mid-Day Meal Programme which is quite successful, should be started wherever it has not been done. While this programme involves additional responsibility on the school management/staff, it is a necessity in the existing situation in our country.

Although this scheme is already in place in various parts of the country, it is being executed with varying degree of responsibility being borne by the organisers. Only 35 per cent (As on 1st April 2003) of the children in primary classes are covered by cooked mid-day meals. The scheme should be implemented in all parts of the country in its true spirit. The *Apex Court* has already fixed January 2005 as a deadline for all states to provide cooked meals to students upto class V.

There are some reports of misuse of the scheme such as showing more students on the role than actually are there. This is just to draw more ration. Such loopholes are required to be plugged by way of regular inspections. Notwithstanding such exceptions we should not miss the central fact that the scheme has been found to be extremely popular in the rural area.

14. Setting up Part-Time and Non-Formal Schools

The problem of enrolment of large number of out of school children could not be solved through formal educational schools established so far. The problem is more acute in the case of enrolment in upper-primary schools. Some children are engaged in part-time work. It has therefore become necessary to consider providing alternate channels of education. It is in this context that the part-time and non-formal schools are being set up.

There is however a danger about this non-formal education. While there will be some sort of improvement in basic literacy, the quality of education in non-formal schools will not be the same as in formal schools. Non-formal education can therefore be used as a transitional phase and focus has to remain on opening more new formal schools and motivating out-of-school children for enrolment in these schools. Further, efforts will have to he made to draw gradually, wherever possible, children from non-formal school to formal schools so as to bring them in the mainstream of education.

15. Special Efforts for Education of Girls

Educated girls will be better able to take care of their families, follow norms of small family, make their children well educated and inculcate better values in their children. There are however some social barriers coming in the way of educating girl child which can be largely addressed by locating schools in the vicinity of their habitation and bringing awareness about the long term advantages of educating the girl child. Psychological barriers can be addressed by establishing wherever possible schools exclusively for girls. Further, it would be better to appoint 75 per cent of teachers in co-educational schools by female and all teachers in exclusively girls' schools by female teachers. Further there are special needs of girl children in terms of availability of infrastructure like toilets exclusively for them.

The enrolment of girls in upper-primary schools will remain a difficulty and this can only be improved either by providing the schools near to their habitation or proper public transport facility.Taking note of low level of enrolment of girl, govt. of India has approved Kasturba Gandhi Balika Vidyalaya Scheme which aims at providing elementary education to the girls child from scheduled castes and tribes, other backward classes and minorities, who will have 75 per cent reservation in the enrolment.

16. Special Efforts for Education of Scheduled Castes (SCs) and Scheduled Tribes (STs)

SCs and STs have suffered centuries of social and economic deprivation and in turn this affected their educational status. The Constitution of India therefore provides for promoting, with special care, educational interests of these groups. There is a positive impact of various special schemes started for these groups. Awareness of the qualitative change in their life is also evident in terms of self-worth and capability that education brings. However there are still some gaps in participation of these groups in education. The enrolment of children (age group of 6-14 years) of these groups especially girls has not been to the level as desired. Poverty continues be the main hindrance. The children are still being used by some of the parents either to look after their younger children at home or help them in their avocation. While efforts will have to be made more vigorously for their participation in education, a larger question of alleviation of poverty will have to be addressed to get most desirable results.

17. Empowerment of Rural Economy

Improvement in the rate of literacy and continuing education, is very much linked with the economic standard of the population. Irrespective of the section they may belong to, the prime concern of the poor families is providing for sustenance of life first. Some families find even meeting day-to-day expenditure so difficult that they prefer their children to help family assignment rather than sending them to school. Better results in education can therefore be obtained if the rural economy in general is improved.

18. Indian Education Through Satellite

The launch of India's first satellite (EDUSAT) in October 2004 dedicated exclusively to education services holds out immense potential for the country. EDUSAT could revolutionise rural education. If five such satellites can be sent up, then a staggering 37 million children in remote areas could benefit at a fraction of the cost involved in primary education now. Thus with the satellite, literacy levels can take a quantum leap forward. However the test of this project lies in determining as to what extent the services of the satellite are actually offered for the primary education in rural area.

Also every state has to invest in establishing state level hubs and infrastructural facilities at the block levels so that the learning process could be revolutionised.

13. ADEQUATE ALLOCATION OF FINANCIAL RESOURCES TO BASIC EDUCATION AND ITS PROPER UTILIZATION

Education costs money, but then so does ignorance—Claus Moser

1. Adequate Allocation to Basic Education

The non-availability of requisite funds is by far the greatest obstacle in maintaining as well as spreading the basic education. On the recommendations of the Education Commission (1966), the Govt. of India (1968) quantitatively accepted to target the investment of 6 per cent of national income in education. Although the allocation for education is steadily increasing in absolute term, the same is decreasing as a percentage of GDP and the same in 2003-04 is 3.74 (BE) and 3.0 (RE) per cent of GDP. To meet the goal of education for all, it is necessary to increase the allocation by half per cent per year and reach 6 per cent of GDP within next 4 to 5 years.

Within the education sector, the Table 2.8 shows the spending on education per student in India and China on primary, secondary and tertiary levels.

Table 2.8: Level wise Spending on Education in India and China

	Public Expenditure per Student			
	Primary	*Secondary*	*Tertiary*	*Sec + Ter*
	(1)	(2)	(3)	(2 + 3)
India	7.2	23.1	92.5	115.6
China	6.1	12.1	85.8	97.9

Source: India Sustaining Reform–2003

The allocation for education in 2005-06 (BE) is 18,337 crore and the allocation to Sarva Shiksha Abhiyan is Rs. 7,156 crore for the year.

It is seen from the above table that although spending by both India and China on-primary level is almost the same, that on secondary plus tertiary level in India is much more than in China. It is pertinent to note that China is already much ahead of India in the literacy rate and therefore India needs to focus the spending within the education sector on primary level vis-a-vis secondary plus tertiary levels. While public expenditure on elementary education should be increased, private entrepreneurs should be encouraged to invest in secondary and tertiary levels.

Table 2.9: Shows the Plan Expenditure on Different Stages of Education in Eighth, Ninth and Tenth Plan.

Table 2.9: Plan Expenditure on Different Stages of Education

Rs. Crore

Education	*VIII Plan 1992-97**		*IX Plan 1997-02 (Central Sector)*		*IX Plan 1997-02 (Central Sector)*		*X Plan 2002-07 (Central Sector)*	
	Outlay	*Per cent*	*Outlay*	*Per cent*	*Expen-diture*	*Per cent*	*Outlay*	*Per cent*
Elementary Edn.	10394	48	16370	66	14523	65.7	28750	65.6
Secondary Edn.	5231	24	2604	10	2323	10.5	4325	9.9
Adult Edn.	1142	5	630	3	520	2.4	1250	2.9
Higher Edn.	2094	10	2500	10	2271	10.3	4177	9.5
Others	740	3	431	2	349	1.6	624	1.4
Technical Edn.	2199	10	2374	9	2110	9.5	4700	10.7
Total	21800	100	24908	100	22096	100	43825	100

* Includes share of States

Source: India 2005 (GOI Publication)

In X Plan, the outlay on elementary education accounts for 65.6 per cent of the outlay on the entire education sector. The funds allocated for primary and upper-primary education in absolute terms are far too inadequate taking into account the requirement to achieve the goal of providing Education For All. Similar is the case of funds for Adult Education which is 2.9 per cent of the education sector. Since elementary and adult education deserve top priority, these should be allotted adequate funds, may be by reallocating funds within the sector as well as between different sectors.

(i) Taking into account the goal of Education for All and the objective of Sarva Shiksha Abhiyan, the allocation to the elementary education has to be suitably enhanced. When the resources are limited there is a greater need to prioritise funding to different sectors and education certainly deserves better treatment. If however inter-sectoral and intra-sectoral adjustments do not bring the allocation for the elementary education to the required level, additional financial resources may have to be raised through levying education cess on the central taxes, as done in the budget 2004-05 and was done in special circumstances like levying surcharge after earthquake in Gujarat etc. in earlier budgets. However while doing so, some of the important points will have to be taken care of like:

(ii) The allocation of cess should be tied up with specific scheme(s) of elementary education and detailed account of its utilization should be maintained.

(iii) The allocation of cess to specific scheme(s) should not lead to reduction in the allocation which would have been a rightful share of such scheme(s) in normal course, meaning thereby, these schemes should not become a conduit for indirectly financing programmes/sectors other than the intended ones.

(iv) Besides Basic Education, Primary Healthcare is another priority area which is suffering for want of funds. Therefore some specific scheme(s) like opening of new Primary Health Centres/Sub-Centres, providing essential equipments at PHC/S-Cs etc. should be clubbed with the elementary education and mid-day meal schemes and requirement of cess should be worked out accordingly.

Incidentally, foreign donors like World Bank, European Union etc also provide some grants to augment the indigenous availability of funds for the literacy drive.

2. Proper Utiiization of Budget Allocation

It is perplexing to note that in many cases it is found that, the concerned education departments do not even utilise the sanctioned funds fully. The main reason for such a state of affairs is archaic procedures adopted by the bureaucracy in implementing various schemes of elementary education. In order to avoid such situations there is a need to regularly monitor the utilisation of sanctioned funds at the appropriate level and take remedial action in case of any shortfall in utilisation.

Another important point is that in many cases the allocated funds are not utilised properly. Any input of resources especially finance should result in improvement in the quality of education apart from its outreach. But this does not always happen. The selection of right kind of teachers and their proper training are the most important factors for improving the quality of education. Rather than spending money on this, sizable portion of it is spent on maintaining burgeoning bureaucracy.

However the above points do not underrate the genuine necessity of increasing the allocation of funds for basic education.

14. CONCLUSION

India began its journey towards providing universal and free basic education with the incorporation of Article 45 in the Constitution stating, "The State shall endeavour to provide, within a period of ten years from the commencement of the Constitution, for free and compulsory education for all children until they complete the age of fourteen years". Its will is further reflected in passing Free and Compulsory Education to Children Bill 2003 making elementary education a Fundamental Right for children in the age group of 6-14 years. There has been a progress in this direction but even after more than 57 years of Independence, the overall literacy rate (1999-2000) in rural area is only 60 per cent and the female literacy rate is 47 per cent. Looking to the population of children to be covered in rural area, there is no doubt that it is a great task. It is not only education in quantitative terms that matters but also education in qualitative terms. The progress achieved so far towards the goal of providing basic education for all is not small. However looking to the stupendous task ahead, there is a need to adopt certain strategies as stated above. For providing necessary infrastructure, facilities and organisation for implementing policies and programmes, one of the major constraints is identified as non-availability of adequate financial resources. This factor has been well recognized and additional funds are required to be allocated especially for basic education schemes. Expending necessary resources are worthwhile as our real wealth is educating each and every child and in future the pay back of higher investment in education is bound to be multifold. Nobel laureate Amartya Sen has rightly said that an investment of 5 paise in elementary education will raise the GDP by 25 paise which means the return is many times more than the investment. However there is a caution and that is - allocation of funds alone will not serve the purpose unless these are effectively utilised.

We must take a cue from China as in the field of basic education the performance of China is better than India although they had more children to be educated.

3

Agriculture and Rural Development

1. INTRODUCTION – STATUS OF LIVING CONDITIONS IN RURAL INDIA

Agriculture is India's lifeline as 65-70 per cent of India's population depends upon agriculture for their livelihood. As per Food and Agriculture Organisation (FAO)'s report on, "Food Insecurity in the World, 2004", the number of unnourished people in India are 221.1 million (2000-2002). Agriculture and Rural Development thus assumes priority in relation to other sectors in our country.

The fast pace in population shift from rural to urban area indicates increasing disparity in the living conditions as well as availability of opportunities for betterment of future between the two societies. This is clearly an indicator of the imbalance in the development that has actually taken place at the ground level whatever may be the policies adopted by the Government. While certain level of shift is advantageous and desirable, reasons for the large-scale shift in population require thorough study and measures for their remedy call for urgent attention.

The major problems being faced by the rural population are abject poverty (leaving aside few people possessing large land holdings), lack of proper facilities of housing, education, healthcare, year round employment and access roads to nearest towns and market centres.

In India, the 'Poverty Line' is defined as expenditure required for a daily calorie intake of 2100 in rural area (2400 in urban area). This is equivalent to half a kg of rice and in monetary terms this is equivalent to Rs. 7-8 per person per day. This expenditure is estimated at Rs. 181. 50 per capita per month in rural area and Rs. 209.50 in urban area, Even with such a low measure, 26 per cent of the total population is falling in BPL category, more than 75 per cent of which live in rural area. This indicates existance of poor standard of living in rural area. At international level, two dollars per day per person is poverty and one dollar per day per person is chronic poverty. If this yardstick is applied to India (1990-2002), 80 per cent of our people qualify as poor and the BPL 35 per cent qualify as worse than chronically poor. In rural area, the incidence of poverty is highest among agricultural labourers, casual workers and small and marginal farmers. It is also more pronounced in socially backward classes such as scheduled castes and scheduled tribes.

As per National Sample Survey 55th round (July 99-June 2000), the literacy percentage in rural area is 60 per cent as against 81 per cent in urban area. In rural area, female literacy is just 47 per cent as against 73 per cent in urban area. Further in rural area itself female literacy (47%) is lesser than males (72%) by 25%. This is an indicator of the slower progress in the field of universal elementary education especially in rural area. As regards healthcare in rural area, the status of important indicators (2001) is given below:

Description	*in Rural area*	*in Urban area*
Crude birth rate per 1000 population	27.1	20.3
Crude death rate per 1000 population	9.1	6.3
Infant mortality rate per 1000 live births	72	42

Source: Statistical Abstract India 2003

The above indicators show the poor medical including maternity facilities available in the rural area. Large number of mothers and their children suffer from malnutrition even as statistics show that about 50 to 80 per cent of rural women are suffering from anemia.

The consumer expenditure per person is much lower in rural area than in urban area, which indicates poor diet taken by the rural population although they are the people who produce the food. The biggest contradiction in Indian agriculture is that despite excess production, the rural masses are not benefited. Poor purchasing power of rural people is actually the reason behind this phenomenon.

Clothing of rural people especially the children is generally poor.

As regards housing in rural area, the requirement of new housing (X Plan 1997-2002) was estimated as 7.7 million units and requirement of housing upgradation was estimated as 20 million units. The thrust of construction activity in rural area needs priority as houses in rural area are generally in dilapidated conditions and it requires special efforts to move construction material there. *Lionel Curtis* once described Indian villages as "*a collection of unsanitary dwellings constructed on dunghill*"

Electricity is yet to reach 16.2 per cent (31.03.2003) of the villages in India. This percentage will increase with the new definition of electrification of a village i.e. a village will be declared electrified only if at least 10 per cent of houses therein are given electric supply. The progress of rural electrification especially in northeastern region is all the more slow compared to other regions.

Notwithstanding the efforts made over the years, 40 per cent of the habitations are still without road connectivity. Even where connectivity has been provided, the roads constructed are not all-weather roads. The conditions are pretty bad especially in hilly areas and states of Bihar, Orissa and Madhya Pradesh. Unavailability of roads or lack of proper roads act as a major hindrance for accessing better facilities like education, healthcare and markets available in towns situated nearby. Only about 90 per cent of rural habitations have been covered with some kind of drinking water facilities. There are many habitations where clean drinking water is not available within a reasonable distance and this is a source of epidemics of gastro-enteritis and cholera cases.

78.1 per cent of households in rural area are without latrines as a result of which many sit for attending to nature's call by the side of the road, which reflects poorly about our country.

2. MAJOR PROBLEMS IN RURAL AREA

Before we dwell on ways and means for rural development, it is necessary to identify major problems being faced by the rural population. These are:

1. Lack of Infrastructure

Road communication is very poor in rural area. There are many villages which are not connected even by *kachha* roads. Often no metal or all-weatherproof roads are available in rural area. State Road. Transport buses do ply on certain routes. However many of the villages do not get covered by these routes. One has to walk down a long distance to board a bus. Due to non-availability of roads or poor condition of the roads wherever they are available, access to marketing facilities available in towns becomes a difficulty. Although telephones are being installed in rural area, many villages do not have this facility. Similarly only a small percentage of villages is having any kind of irrigation facility. Also agriculturists find great difficulty in getting uninterrupted power supply to run their pump sets for irrigation. Thus there is a lack of even bare minimum rural infrastructure facilities in most part of the rural area.

2. Lack of Basic Facilities

There are number of villages in which basic facilities like elementary education, clean drinking water, health services, sanitation are not available. In some villages there are schools but not enough classrooms are available. In some cases villagers have to fetch water from more than I km distance. Access to clean drinking water is available to the extent of 73.2 per cent. Access to even primary healthcare is difficult. It is not unusual scene to see that sick persons are taken to the nearest Primary Health Centre (PHC) that is situated at a distance of 5 to 20 km by a bullock cart.

3. Large Family Size and Fragmented Land Holdings

Rural population largly depends upon agriculture. Poor response to the concept of adopting small family norm in rural area has resulted in fragmentation of land holding with each passing generation. Without irrigation facility these small units are becoming mostly unviable. This is a major cause for increasing poverty in the rural area and more than 27 per cent of the rural population lives below the poverty line.

4. Unemployment

Cottage industries in rural area could not thrive on account of tough competition from mechanised industries. The cost of items produced by the industries is bound to be cheaper as they employ machines of mass production. In absence of growth of cottage industries, maximum load of employment in rural area falls upon agriculture. Since employment potential in agriculture has limitations and the work in agriculture is of seasonal nature, the rate of unemployment in rural area is rising at a faster pace.

5. Low Labour Productivity and Low-level of Mechanisation

The social relationship in the rural area is undergoing a change in view of economic pressures. This has also affected the bond of relationship that existed earlier between the employer and the worker and consequently resulted in the lower labour productivity. Application of technology and large scale mechanisation in agricultural operations can substantially improve the productivity but this could not happen in a big way due to steady decrease in the size of land holdings over a period of time.

6. Indebtedness and Distress Sale of Produce

Due to low income from the agricultural produce and high expenditure on the inputs coupled with considerable family expenses which include expenditure on various ceremonies like birth, marriage and death, most of the farmers inherit the burden of loan and its interest. The interest rate of the banks' loan so far used to be more than 12 per cent which has since been redeuced to 8½ to 9 per cent. Under normal credit schemes, during the current year, loan can be availed only when the past years' loan is fully paid. Due to this pressure as well as absence of facility of safekeeping the produce, farmers are compelled to sale their agricultural produce immediately as soon as it is ready in the fields. At the time new produce comes in the market, the prices are the lowest. Sale of their hard won precious produce at this point of time thus amounts to a "distress sale".

7. Poor Marketing Infrastructure

It is generally seen that the prices go up by at least 20 per cent to 25 per cent well past 6 to 8 months of arrival of new produce in the market. However due to distress sale by the farmers, middlemen and wholesale purchasers make handsome profit without going through the ordeal which ordinary farmer undergoes throughout the year. There are some State sponsored well intentioned procurement schemes like cotton procurement scheme in Maharashtra. These schemes, however, could not serve the real interest of the farmers at number of places due to mismanagement and vested interests both by the management as well as staff administering the schemes. Further under these schemes money is not given to the farmers in one go but in two or three stages. Most of the farmers are not in a position to wait that long and therefore the advantage of such schemes is taken by a small cross section of the farmers. These schemes, however, serve positive role as additional window and exert some check on the private purchasers.

8. Non-availability of Quality Seeds in Time

Agriculture faces major impediment due to non availability of quality seeds of improved varieties, and particularly at a time when it is needed. Although it is known in advance when the monsoon is going to arrive, initially there is always delay in availability of seeds in the market thus creating panic in the farming community leading to spree for buying them even if the seeds are of inferior quality.

9. Absence of Budgeting Concept and Investment

Large section of urban population is having assured monthly income and they plan their expenditure accordingly. On the other hand income from the agriculture depends upon vagaries of nature i.e. rain etc. It also depends upon unforeseen factors such as incidence of damage due to pests etc. Traditionally rural people have strong adherence to social customs and besides their regular family expenses, substantial part of their earnings is spent on social functions such as birth, marriage and death. On many occasions they even take loan to conduct these functions. There is no concept like budgeting for the month or the year like in urban families. Thus at the end of the year, no money is left for cultivation leave alone saving money for investment for the long term development of the agriculture or for a social security for old age. Therefore for cultivation, they have to avail loan facility which is extended by the State cooperative banks or seek loan from the private lenders. Thus practically there is not only no improvement in the financial status but many of them land into more debt burden on account of lower income from the produce and higher input costs.

3. PRE-INDEPENDENCE AND POST-INDEPENDENCE APPROACHES TO RURAL DEVELOPMENT

1. Basics of Rural Development

1.1. Social Consciousness and Awareness

For any development to take place, there should be realisation about the need for it. Further, for sustenance of the development, the real effort has to come from within. Unless it is there, any amount of effort from outside will not be wholly successful. The rural masses are habituated with following age-old social, occupational and economic practices. These need to be changed to suit the fast changing environment. For example, ways and means have to be found to improve productivity in agriculture. Elsewhere in other countries a lot of technological advancements are being utilised in agriculture. The awareness about the need to change can be best brought through educating them about their strengths and weaknesses, rights and duties. Besides the State Govt. and institutions like Panchayat Raj, dedicated NGOs can play important role in bringing awareness in the rural masses.

1.2. Village Leadership

This is one of the most important aspects of the development process. If the leadership is dedicated to the cause, the process becomes easier. Since village is a small unit of the society, it is necessary that the village leadership focuses on development of the village irrespective of the political party affiliations, and villagers also should choose representatives who can contribute to development of the village. It is also moral responsibility of political parties to keep interest of the rural poor and uneducated villagers above party lines.

1.3. Balanced Approach for Rural-Urban Development

Rural population provides basic necessities of life in the form of food, fibre and raw material to urban population. In turn urban population thrives on trading and marketing of these products apart from industrial activity in their area. Similarly rural population depends upon several organised facilities available in towns and cities like higher education, hospitals, marketing etc. The development of these two sectors is thus interdependent and therefore development of both is equally important. However at ground level there is one basic difference. Since there is more awareness in urban population about the need of development and there is an organised effort to attain it, the pace of development is faster and easier in urban area. On the other hand due to abject poverty and lack of education, rural population on its own is not in a position to accomplish proper development. It needs outside help from the State or the NGOs, besides efforts from their own side to improve the situation.

2. Early Approaches to Rural Development

Having identified problem areas as mentioned, strategies for their solution are to be found which are best suited to socio-political-economic system obtained in rural area. During freedom struggle and immediately later, our great national leaders studied these problems in depth and worked hard for bringing reforms in the systems existing then. The approaches of various leaders might have differed but the goal was the same i.e. development of the society. Gandhian ideology includes most of the basic elements suggested in these approaches and the same are discussed here.

2.1. Gandhian Ideology on Village/Rural Development

Gandhiji held very strong view about the importance of village on the national canvas. His idea of village was that it should be self contained unit, largely not dependent on its affairs to be ruled from outside. He believed in raising the potential of the rural population so as to be self reliant and govern and to be governed by themselves. He expressed his views as:

"Just as universe is contained in the self, so is India contained in the villages. India lives in villages. If the village perishes, India will perish too." "My idea on village *Swaraj* is that it is a complete republic, independent of its neighbours for its own vital wants, and yet inter-dependent for many others (wants) in which dependence is a necessity. The government of the village will be conducted by *Panchayat* of five persons annually elected by the adult villagers, male and female, possessing minimum prescribed qualifications. These will have all the authority and jurisdiction required. Since there will be no system of punishments in the accepted sense, this *Panchayat* will be legislature, judiciary and executive combined to operate for its year of office."

"Lionel Curtis describes our villages as dung heaps. We have to turn them into model villages. Our village-folk do not get fresh air, though they are surrounded by fresh air; they don't get fresh food, though they are surrounded by the freshest foods."

"When our villages are fully developed, there will be no dearth in them of men with a high degree of skill and artistic talent There will be village poets, village artists, village architects, linguists, and research workers. In short, there will be nothing in life worth having, which will not be had in the villages."

"I know village economics. I tell you that the pressure from the top crushes those at the bottom. All that is necessary is to get off their backs."

"The idea behind the village industries scheme is that we should look to the villages for the supply of our daily needs, and that when we find that some needs are not so supplied, we should see whether, with a little trouble and organisation, they cannot be profitably supplied by the villagers. In estimating the profit, we should think of the villager, not of ourselves."

"If he (village worker) will go to the village as a teacher, he will go there no less as a learner. He will soon find that he has much to learn from the simple villagers."

Gandhiji found in a thread, coming out of a spinning wheel (*Charkha*), a force binding millions of countrymen, poor or rich, together. He said, "Spinning for me is an emblem of fellowship with the poorest of the land, and its daily practice is a renewal of the bond between them and ourselves."

Gandhiji strongly propagated the idea of developing village industries because they could be started with in-house raw materials produced by the villagers themselves and these would not call for high capital. At the same time he was not opposed to industrialisation but he wanted it for the objective of manufacturing machinery for helping villages to enhance their production capability. Gandhiji disliked monopoly and anything of which benefit cannot be shared with the masses.

2.2. Ideologies of Gandhiji and Karl Marx

Both Gandhiji (1869-1948) and Karl Marx (1818-1880) championed the cause of poor and downtrodden. Both advocated new socio-political-economic order in which the common people would enjoy full human rights without exploitation by the privileged class. Gandhiji as such was not against wealthy but he expected them to use only that portion of wealth which is sufficient to fulfil their need and for the rest they should act as trustee. Both opposed monopolistic nature of economy.

There were however basic differences of approach for achieving the objectives. Gandhiji and Vinoba Bhave stood for Sarvodaya. While Gandhiji believed in persuasive approach in bringing about a new social order, Karl Marx believed that the privileged class would never like to part with their wealth willingly and therefore justified the use of coercion and force against the rich if necessary. This philosophy of Marx amounted to justification of class war in the pursuit of achieving the objective. For Marx, cause was more important than the means to achieve the cause. For Gandhiji, means were as important as the cause.

2.3. Important Rural Reconstruction Programmes/Schemes taken-up

Before independence, there were several reformers who were fired with the spirit of public service. They started various innovative schemes aimed at rural development. Notable of these are:

2.3.1 Sriniketan Rural Reconstruction Programme (1921)

Rabindranath Tagore set up Ashram at Shantiniketan (West Bengal), a peaceful place away from the city life of Calcutta. He started the rural reconstruction programme from there. The programme emphasised on helping villagers to develop agriculture, livestock, formation of cooperatives and improving village sanitation etc. One of the objectives of the programme was to associate the youth in the Ashram with the rural reconstruction work.

2.3.2 Martandam Rural Reconstruction Programme (1921)

The programme was started at Martandam in South Travancore by Spencer Hatch of Y.M.C.A. The programme emphasised on betterment of living conditions of the rural poor through constructive schemes of the Y.M.C.A. The programme covered development of cottage industries like mats and basket making, poultry keeping, bee-keeping etc.

2.3.3 Gurgaon Scheme

The Village Development Programme was started by F.L. Brayne, a Britisher, in Gurgaon district in Punjab & Haryana. The programme included establishment of School of Rural Economy to train the village guides for rural uplift work; setting up Domestic School of Economics to train the village women; taking up rural sanitation works; agricultural development programmes, etc. He especially visualised the role of village school teacher as beyond teaching in the school and as a person who can solve the simple problems of the villagers.

2.3.4 Rural Reconstruction Programme (1890) in Baroda

The rural reconstruction programme was started in Baroda province by Maharaja Sayaji Rao. The programme emphasised on establishment of village panchayats; education of children; adult education; improving agricultural production through expansion of irrigation facilities; production, disιribution and use of better quality seeds, etc.

2.3.5 Firka (Village) Development Scheme (1946)

Firka Development scheme was first started in 34 firkas in the then Madras State by T. Prakasam. The scheme was later extended to other firkas. Based on Gandhiji's idea of "village swaraj", the programme emphasised on providing housing, education, water supply, development of roads, improvement of sanitation etc. At macro level the programme emphsised on improvements in agriculture, development of village industries, formation of panchayats, setting up co-operatives, encouraging social and cultural activities etc. The scheme was later merged in Community Development Programme in 1952.

2.3.6 Etawah Pilot Project (1948)

The project was started at Mahewa in district Etawah (UP) by M. R. Albert Mayer who was an American town planner and remained in India during 2nd World War. The project emphasised on training of village level workers, outside job training, training to village leaders, social service workers, adult literacy teachers, functionaries of panchayat etc. It also emphasised on active participation of the villagers in various programmes, development of agriculture and village organisations such as panchayats, schools and co-operatives etc.

2.3.7 Nilokheri Project (1948)

The Project was started by S.K. Dey to rehabilitate the displaced persons coming from Pakistan (due to partition of the country) by establishing a new township at Nilokheri in Karnal district in Hariyana. Dey engaged the refugees in constructive work assuring them the right to live, the right to work and right to earn remuneration for the work they performed. These three rights formed the basis of a scheme known as *Mazdoor Manzil*. The plan included developing medical facilities, sanitation, middle and high school education, vocational training, cottage industries, agricultural extension covering farming, horticulture, poultry, fishery, etc.

All the above programmes contained essential features of the programmes initiated by Gandhiji from Sabarmati and Sewagram Ashrams set up by him. The programmes were based on the principle of rural reconstruction through the collective efforts of the villagers themselves. These included development of communications (roads), rural water-supply, education, adult literacy, improvement of sanitation, improvements in agricultural practices, developing participative approach and village leadership and developing institutions like co-operatives etc. Through these programmes, awareness about the evils of child marriages and other undesirable social customs and practices was also spread in the villages.

3. Post-Independence-Rural Development as a Part of the Planning Process

After independence, rural development formed important part of the planning process both at the National and State level. The schemes were formulated after deliberations in various forums, involving the concerned in the planning process. The schemes generally fell in two major groups: Schemes in which area approach was adopted and Schemes which focused on certain target groups. These schemes over a period of time and depending upon the emphasis and importance given by the Government in power underwent changes both in terms of their names as well as content. The important schemes which are presently in vogue are narrated below:

3.1. Pradhan Mantri Gram Sadak Yojana (PMGSY)

Pradhan Mantri Gram Sadak Yojana, a fully Centrally sponsored scheme, has been launched on 25th December, 2000 with a view to provide connectivity through all weather roads to all habitations. The objective of the scheme is to cover habitations with a population of

1000 persons and above in three years and all unconnected habitations with a population of 500 persons and above by 2007. In respect of the hill states and desert areas, the objective would be to connect habitations with a population of 250 persons and above.

3.2. Swarnajayanti Gram Swarozgar Yojana (SGSY)

This scheme has been launched on 1st April, 1999 after merging into it the programmes of Integrated Rural Development Programme (IRDP), Training of Rural Youth for Self-Employment (TRYSEM), Supply of Toollkits to Rural Artisans (SITRA), Development of Women and Children in Rural Areas (DWCRA), Million Well Scheme (MWS) etc. The objective of SGSY is to assist rural families below poverty line to take up self income generating schemes by providing them bank credit and Govt. subsidy for starting the schemes. It provides for involvement of NGOs/Banks/Individuals as facilitator/Self-Help Promoting Institutions in nurturing and development of Self-Help Groups (SHGs) including skill development. The scheme provides for skill development training based on local requirement.

The focus is on establishing a large number of micro enterprises in rural areas based on the ability of the poor and potential of each area for a sustainable income generation. The subsidy allowed is 30 per cent of the total project cost, subject to a ceiling of Rs. 7500 (for SC/ST). For SHGs, subsidy is 50 per cent of the project cost subject to a ceiling of Rs. 1.25 lakh or per member subsidy of Rs. 10,000, whichever is less. SHGs may consist of 10-20 members. There is no monetary ceiling on subsidy for minor irrigation projects for SHGs as well as individual swarozgaris. SCs/STs would account for at least 50 per cent of the swarozgaris and women for 40 per cent.

The SGSY is being implemented through District Rural Development Agencies (DRDAs), with involvement of Panchayat Raj Institutions, Banks and the NGOs. Since inception of the programme, 19.69 lakh SHGs have been formed covering 45.67 lakh swarozgaris. These include 20.13 lakh members of the SHGs and 25.54 lakh individual swarozgaris.

3.3. Sampoorna Grameen Rozgar Yojana (SGRY)

This scheme has been launched on 25th September, 2001 by

merging the Employment Assurance Scheme (EAS) and Jawahar Gram Samridhi Yojana (JGSY). JGSY was itself reconstructed from the earlier scheme named Jawahar Rozgar Yojana (JRY) created by merging National Rural Employment Programme (NREP) and Rural Landless Employment Guarantee Programme (RLEGP). Beside other provisions, under Sampoorna Grameen Rozgar Yojana, free food grains are offered to States to take up Food for Work Programme (FWP).

The objective of the programme is to provide additional wage employment in rural areas as also food security, alongside creation of a durable community, social and economic assets and infrastructure development in these areas. It is executed through two streams, viz., (i) First Stream replaces the EAS, which is implemented at the District and Intermediate Panchayat levels. 50 per cent of the SGRY fund is earmarked and distributed between the Zilla Parishads (20 per cent) and the intermediate level Panchayats or Panchayat Samities (30 per cent); (ii) Second Stream replaces JGSY and is implemented at the village Panchayat level. 50 per cent of the SGRY fund is earmarked for this stream. The fund is released to the Village Panchayats through the DRDAs/ Zilla Parishads.

The scheme envisages providing 100 crore mandays of wage employment every year. The programme is being implemented on a cost sharing basis between the Centre and the State in the ratio of 75:25. The Centre provides its share in the form of foodgrains. The Gram Panchayats can take up any work with the approval of Gram Sabha as per their felt need and within available funds. 50 per cent of the funds earmarked to the Gram Panchayats are to be utilised for infrastructure development works in SC/ST localities. No contractors or intermediate agencies can be engaged for executing work under this scheme. Five kg of foodgrains will be made available per manday to the workers and the balance of wages will be paid in cash. The overall supervision of the programme rests with Zilla Parishads. The programme will be evaluated through impact studies conducted by reputed institutions/ organizations sponsored by the Central/State Govts.

As mentioned earlier, the NREP and RLEGP have been merged in SGRY. The background of initiating these important programmes and their salient features are:

3.4. National Food for Work Programme (NFWP)

Agriculturists are quite busy during 4 months and there is slackness in employment during remaining part of the year. It therefore becomes difficult for the poor to sustain the living during slack season. Food for Work Programme was started in 1977 to take care of this particular aspect and at the same time to create in rural area, durable productive capital assets like roads, bridges, minor irrigation projects-construction of tanks/ponds/bunds, canals etc. The workers under this programme were paid partly in cash and partly in foodgrains by utilising the surplus stocks of foodgrains. Initial response to this scheme was good but due to certain shortcomings, response started dwindling. In view of this the Food for Work Programme was reshaped into a broad based programme called National Rural Employment Programme in 1980.

National Rural Employment Programme (NREP): The objective of the programme is to provide employment to rural workers particularly at a time when they are not able to find gainful employment, and at the same time to create durable community assets thus strengthening rural infrastructure.

The salient features of the programme include: (i) District/ Block level employment plan will be formulated, (ii) Priority is given to works like soil and water conservation, irrigation, flood control and drainage, construction and improvement of tanks/ponds, social forestry and pasture development, school and dispensary buildings, works relating to hygiene and sanitation etc., (iii) 10 per cent of the resources are earmarked for works of benefit for SCs/ STs, (iv) The execution of the work will be through Panchayat Raj Institutions and implementation has been entrusted to the District Rural Development Agencies. It is a Centrally sponsored scheme on 50:50 shared basis between the Centre and the State. The Centre will provide its share in the form of foodgrains, (v) NGOs can also be entrusted with the execution of the work.

With a view to improve the nutritional level, the foodgrains are to be provided at the rate of 5 kg per manday.

Rural Landless Employment Guarantee Programme (RLEGP): Rural Landless Employment Guarantee Programme was started in 1983. The difference between NREP and RLEGP is that while NREP is for generation of additional employment for the unemployed and underemployed persons, RLEGP provides guaranteed employment to at least one person of every landless household up to 100 days in a year.

Food for Work Programme has since been redesigned in Nov. 2004 and renamed as National Food for Work Programme (NFWP).

3.5. Jai Prakash Rozgar Guarantee Yojana (JPRGY)

This scheme is intended to provide employment guarantee to the unemployed in the most distressed districts of the country.

3.6. Antyodaya/Antyodaya Anna Yojana

The Antyodaya scheme is intended for the upliftment of the poorest of the poor. The selection of beneficiaries under this programme is solely based on the extent of the poverty of the family and targets the families below the poverty line. These families are provided with 35 kg of foodgrains per family per month at a highly subsidised price of Rs. 2 per kg of wheat and Rs. 3 per kg of rice.

3.7. Minimum Needs Programme (MNP)

The objective of Minimum Needs Programme, started in 1974, is to define minimum needs so as to improve the quality of life of the rural population and provide basic infrastructure for supporting the various rural development schemes taken up for implementation from time to time. The important components of this programme are:

Elementary Education: The basic objective is to achieve universal elementary education to all within the age group of 6-14 years. Since in this age group, girls constitute 88 per cent of all un-enrolled children (in school), special efforts have to be made to encourage them to join the schools. In order to supplement the coverage by the formal education system, non-formal education programme has also been taken up.

Adult Education: The objective is to achieve literacy for all persons in the productive age group of 15-35 years. The success of this programme is dependent upon motivation of the learners as well as finding by them suitable time for learning as most of them are engaged in agriculture or allied activities for earning bread.

Healthcare: The objective is to cover healthcare facilities to all by providing three-tier system comprising district hospital, primary health centre and its sub-centre.

Village Health Guides Scheme has been started to educate the villagers about basic health related matters.

Nutrition: The objective is to provide improvement in the level of nutrition to school going children and others through mid-day meal scheme and Special Nutrition Programme.

3.8. Rural Water Supply Programme

Drinking water facilities to rural habitations are provided under the State Sector programmes. The efforts of the states are supplemented by the Centrally sponsored Accelerated Rural Water Supply Programme (ARWSP). The objective is to provide safe drinking water to all habitations in five years. The programme also includes tackling the problem of water quality in affected habitations and to institutionalise water quality monitoring and surveillance systems and to promote sustainability, both of systems and sources, to ensure continued supply of safe drinking water in 'covered' habitations.

The programme has been revamped by incorporating the following concepts for ensuring peoples' participation: (i) adoption of demand-driven responsive and adaptable approach of villagers to ensure their full participation in the choice of scheme design, control of finances and management, (ii) shifting the role of Govt. from direct service delivery to that of facilitator and (iii) partial cost sharing either in cash or kind or both and full responsibility of operation and maintenance by end users. 67 districts all over the country have been identified for implementing above reforms on a pilot basis.

The reforms initiated as above have been applied to *Swajaldhara* scheme launched countrywide on 25th December, 2002 with a provision of 90 per cent of project cost to be met by the Government. A scheme named *Pradhan Mantri Grameen Jal Samvardhan Yojana* has been started with the main focus on water conservation and management.

3.9. Rural Sanitation Programme

The State looks after the rural sanitation and the efforts of the State are supplemented by the Centre through Central Rural Sanitation Programme (CRSP) launched in 1986. The state wise allocation based programme has since been phased out by March 2002 and Total Sanitation Campaign (TSC) has been introduced. The TSC emphasises the awareness building approach and it includes construction of individual household latrines, toilets for Schools/ Balwadis/Anganwadis and sanitary facilities for women

3.10. Indira Awas Yojana (IAY)

To meet the housing needs of the rural poor, Indira Awas Yojana (IAY) was launched in May 1985. It aims at helping rural people below the poverty line belonging to SCs/STs and non-SC/ST categories in construction of dwelling units and upgradation of existing *kutcha* houses by providing grant-in-aid. A minimum of 60 per cent of the funds are to be utilised for construction of houses for the SC/ST people. Sanitary latrine and smokeless *chulha* are integral to the unit. Selection of the beneficiaries is done by the Gram Sabha. Assistance for construction of new houses is provided at the rate of Rs. 25,000 and 27,500 per unit in the plains and hilly areas respectively and for upgradation *of kutcha* houses at the rate of Rs. 12,500.

Built into IAY is a credit-cum-subsidy scheme for rural households. A subsidy up to Rs. 10,000 and loan up to Rs. 40,000 are provided to eligible households.

In order to complement IAY, the Golden Jubilee Rural Housing Finance Scheme was launched in 1997 to give boost to rural housing. Since inception of IAY, upto March 2004, about 14.78 lakh houses have been constructed.

3.11. Accelerated Irrigation Benefit Programme (AIBP)

This will assist the states to accelerate the completion of unfinished medium and major irrigation projects, and also to undertake reforms by revising user charges and setting up of water users' associations.

3.12. Watershed Development Programme

There are three Centrally sponsored area based watershed programmes for development of wastelands/degraded lands, i.e. Drought Prone Areas Programmes (DPAP), Desert Development Programme (DDP) and Integrated Wastelands Development Programme (IWDP). The DPAP was launched in 1973-74 to tackle the special problems faced by those areas which are constantly affected by drought conditions. The DDP was launched in 1977-78 to mitigate the adverse effects of desertification. The cost of the projects is to be shared by the Centre and the States in the ratio of 75:25. Evaluation part of the projects is now left to the States.

Under an initiative called "Hariyali" launched in January 2003, all the above programmes would be implemented through the Panchayat Raj Institutions. Gram Panchayats shall implement the projects under the overall supervision of Project Implementation Agencies (PIAs). Intermediate Panchayat may be the PIA for all the projects sanctioned to a particular Block/Taluka. In case, these Panchayats are not adequately empowered, then the Zilla Panchayat can either act as a PIA itself or may appoint a suitable Line Department like Agriculture, Forestry etc. Failing these options, the ZP/DRDA may consider appointing a reputed NGO with adequate experience and expertise in the field of work as PIA.

Lok Nayak Jai Prakash Narayan Fund (LNJPNF)/Agriculture Infrastructure and Credit Fund (AICF) to the tune of Rs. 50,000 Cr has been set up in 2004-05. The activities under this fund include minor irrigation, micro/drip/sprinkler irrigation, rainfed agriculture, integrated watershed development, flood control and drainage etc.

3.13. Accelerated Rural Electrification Programme (AREP)

This scheme is intended to accelerate electrification of 80,000 villages that have no access to electricity. Three schemes namely Accelerated Rural Electrification Programme, Kutir Jyoti Yojana and rural electrification component of the Prime Minister's Grameen Yojana have been merged together.

3.14. Integrated Rural Energy Programme (IREP)

The Integrated Rural Energy Programme was launched with the objective of providing energy for meeting the basic needs of cooking, heating and lighting by utilising locally available resources. Apart from energy for various activities, this would result in creation of employment and increase in productivity and incomes.

3.15. Tribal Area Development Programme (TADP)

The tribal people, who constitute 7.4 per cent of the total population of the country and literacy of 16.4 per cent as against 64.84 per cent of the country as a whole, have a distinct culture of their own. Their population is isolated and scattered over a large area. They are also often subjected to displacements when new projects are taken up and adequate attention is not paid for their rehabilitation. It is therefore necessary to make special efforts for their development and bring them into the mainstream. TADP was therefore taken up under the Ministry of Agriculture. The Tribal Sub-Plan (TSP), as a part of TADP, aims at development of socio-economic conditions of tribal population with emphasis on agriculture, livestock, pasture development and literacy. The plan is funded by central and state governments. TSP further identifies cluster of tribal population and bring them under Integrated Tribal Development Projects (ITDP) for focused attention.

3.16. Pradhan Mantri Gramodava Yojana (PMGY)

This scheme has been introduced with a view to allocate additional central assistance to the states in order to achieve the objective of sustainable human development at the village level.

3.17. PURA

It is proposed to start a programme for Providing Urban services to Rural Areas under this scheme.

PURA is aimed at expanding production and employment in the unorganised enterprises around existing clusters of industrial activities and services as well as formation of new clusters so that PURA will offer possible solutions to a number of problems that afflict rural area such as unemployment, isolation from markets, lack of connectivity and migration to cities. It is necessary to identify a few growth poles and take up these as pilot projects.

3.18. Rural Infrastructure Development Fund (RIDF)

The Rural Infrastructure Development Fund was established in NABARD in 1994-95 to promote development of infrastructure in rural area.

A corpus of Rs. 8,000 crore has been provided for this in the budget 2005-06 as in 2004-05.

3.19. Policy on Resettlement and Rehabilitation for Project Affected Families (PAF)

3.19.1. GOI National Policy on R&R for PAF-2003

Among other provisions, the policy provides for: (i) it applies to projects displacing 500 families or more en mass in plain areas and 250 families or more in hilly areas, (ii) Allotment of agricultural land or cultivable wasteland to the extent of actual land loss subject to a maximum of one hectare of irrigated land or two hectares of un-irrigated land/cultivable wasteland subject to availability of Govt. land in the district, (iii) Allotment of house sites free of cost to PAF whose houses have been acquired. The policy also provides financial assistance specifying amounts depending upon the circumstances mentioned therein. The policy also provides for appointments at the state level of an Administrator (not below the rank of Secretary) and staff for R&R in respect of the projects in the states. While there are some positive features of the scheme, the implementation of the policy will involve administrative expenditure on the separate set up as stated above.

3.19.2. National Thermal Power Corporation Policy on R&R - 1993

The NTPC is a leading Central Public Sector Undertaking which has executed number of projects some of which have been funded by the World Bank etc. who are quite particular about framing proper R&R policy and its implementation. The land that is acquired for projects is though for a public purpose, generally the displacement is involuntary. The Project Affected Persons (PAPs) often face eviction and, in fact, have no choice but to face the new social set up. Further, changes in the land use pattern substantially alters the agro-based rural economy and the life style of the people is affected drastically. In such cases the rehabilitation Action Plan (RAP) is formulated so that after a reasonable transition period, the displaced persons improve, or at least regain their previous standard of living, earning capacity and production levels.

To finalise the Rehabilitation Action Plan (RAP), a consultative approach is adopted. For each concerned village, a Village Development Advisory Committee (VDAC) is constituted comprising sarpanch, panch, two land oustees, social worker and upsarpanch. These VDACs in association with the district administration and NTPC play an important role in the preparation and implementation of the Rehabilitation Action Plan. For this purpose Rehabilitation Committee is formed comprising Land Acquisition Officer; representatives of Collector, Zilla Parishad and NTPC; Employment Officer; Labour Officer; Agriculture Officer; and Sociologist.

Project Affected Persons are categorised depending upon the extent of land acquired, homestead acquired, status of owner/tenant of the land, whether the PAP is a tribal person etc. Depending upon category of PAPs, they are offered one of the four sets of entitlements. These entitlements consist of: Set-A: Land for land, or self employment, or shop or award of petty contract, or job; Set-B: Alternate house site measuring 50 ft by 40 ft along with community infrastructural facilities and relocation, and other rehabilitation assistance; Set-C: Self employment or shop or award of petty contract or job; and Set-D: Assistance to establish the type of facilities lost on account of land acquisition. While person loosing entire land or person left with land not economically viable (less than 2 ha) is eligible for Set-A, person whose homestead is acquired is eligible for Set-B and person whose homestead and land is acquired is eligible for Set-A and Set-B.

As per the socio-economic survey, around 80 per cent of the Project Affected Persons whose land is acquired are farmers with small land holdings. Therefore among all the options, Land for Land is the most preferred modes of rehabilitation. However, the main problem is the availability of land. If the revenue land is available nearby the place of resettlement, there is no problem. If

Govt. land is not available, private land may be acquired through a Land Purchase Committee (LPC) on a "willing buyer-willing seller" basis in which case NTPC will use its good offices to facilitate this.

The stress is laid on providing opportunities of self employment. The dairy development, poultry raising, handicraft making etc. are some of the most viable options. NTPC provides Rs. 15,000 per family for this purpose. Technical assistance is also provided to start the venture as well as marketing the products. Those members of the displaced families who wish to gain vocational training, NTPC pays the training cost including tuition, residential hostel charges, books and stationary charges.

Apart from specific schemes for families, community facilities in the resettlement area, which is situated in the vicinity of the project, are provided which include provision of primary school, dispensary, village pond, panchayat ghar, one drinking water well for at least every 50 families, renovation of existing well/handpump, water supply for drinking, water bound macadam (WBM) with bituminous surfacing road linked to the main road, drainage, road inside the village with trees planted on either side, *sulabh souchalaya* for each dwelling area and road electrification.

NTPC is further improving its R&R 1993 policy, which envisages to offer still better facilities. NTPC has also formulated, "NTPC Corporate Social Responsibility-Community Development (CSR-CD) Policy-2004" The salient features of this policy are:

The objective of the Policy is to address the socio-economic issues and continuing the community and peripheral development work in the neighbourhood area of its operating stations. The programmes under this policy include education (skill development, vocational training), self-employment, health (immunization, family planning, medical camps), infrastructure development, conducting rural sports, veterinary camp, forestry and co-operatives.

The example of NTPC is worth emulating by other corporates too.

4. Community Development Programme

The Community Development Programme was started way back in 1952 with the objective of achieving active participation of the rural people in the development process. It is based on the principle that for success of any programme, involvement of the people for whom the scheme is taken up is essential. To an extent this programme became the mother programme for all the rural development programmes to follow, as the principle of participation became the guiding one for all the succeeding programmes.

The Community Development Programme touched development of all aspects of rural life such as education, health, drinking water, sanitation, roads, irrigation, agriculture, cottage industries etc. The programme envisaged contribution from the community to the extent of 50 per cent in the form of providing labour and materials. The objective was to utilise the local labour and materials so that the local people get the employment as well as local material is used and converted into assets for the benefit of local community.

To achieve greater and effective participation of the masses, in a vast country like India, centralisation of the policy and decentralisation of the implementation process is essential. Thus while at the Central level, the Ministry of Rural/Community Development, Panchayat Raj and Cooperation has overall control over these programmes and decides policy matters, the responsibility of actual implementation of different programmes is with the State Governments. At State level there is a State Development Committee of which Chairman is the Chief Minister. Secretary of the Committee named Development Secretary is coordinating and monitoring the implementation of these programmes. At district level, while Zilla Parishad has the overall responsibility of the implementation of the programmes, at block level, the Panchayat Samiti and at the village level, the Gram Panchayat have this responsibility. At the administrative level, Chief Executive Officer (CEO), Block Development Officer (BDO) and Village Development Officer (VDO) are the functionaries at the district, block and village level respectively. The BDO and other officials are drawn from the revenue department. This programme has been extended with the name of National Extension Service.

5. Rural Administrative Structure

In India, District is the important unit of administration. As per 2001 Census, there are a total of 593 districts in the country. The Collector is the administrative in charge of the district. Among others, the collection of revenue and maintenance of law and order are his important functions. A district is further divided into 3 or more subdivisions each headed by Assistant Collector or Sub-Divisional Officer. Each sub-division is again divided into taluks with Tahsildar as in-charge of each taluk. Each taluk may comprise 200 to 600 villages. The Tahsildar is assisted by Niab-Tehsildar and Patwaris. Patwari is the last important revenue post in the rural area. Each Patwari is in-charge of a village or a group of villages depending on the population and the area.

6. Panchayat Raj

Gandhiji advocated a concept of self-governance by each village and for that purpose should form its own Panchayat and the same should govern all the affairs of the village. In this context, based on the recommendations of the Balwantrai Mehta Committee, the three-tier institution of Panchayat Raj (PR) was constituted in 1959. Under this scheme, there is Zilla Parishad at district level, Panchayat Samiti at block level and Gram Panchayat at village level. 73rd Amendment Act, 1992, which came into force on 24th April 1993 further strengthened the institution of PR by incorporating provisions like: Reservation of membership for SC/ST and women not less than 33 per cent; Elections to the institutions after every 5 years, Constitution of District Planning Committee to prepare plan for development of the district; Constitution of State Finance Commission for empowering financial powers of the Panchayats; Providing constitutional status to the Gram Sabha with powers like selection of beneficiaries for certain schemes

According to the Constitution, Panchayats shall be given powers and authority to function as institutions of self-government. The powers and responsibilities to be delegated to Panchayats at the appropriate level are: (i) preparation of a plan for economic development and social justice, (ii) implementation of schemes for economic development and social justice and (iii) to levy, collect and appropriate taxes, duties, tolls and fees.

6.1. Village Level

Panchayat Raj has its foundation in the villages and at the village level it comprises Gram Sabha and Gram Panchayat.

6.2. Gram Sabha

The Constitution (73rd Amendment) Act gives constitutional status to the Gram Sabha. It is a body consisting of persons registered in the electoral rolls relating to a village comprised within the area of Gram Panchayat. Thus all villagers above the age of 18 years have an inherent right to determine their own destiny. It has been envisaged that Gram Sabhas shall plan and prioritise development works to be taken up in the village; approve the annual plan for the Gram Panchayat; seek active participation of women, SCs/STs; ensure transparency in the working of Gram Panchayat; exercise the right to check the accounts of the Gram Panchayat; select beneficiaries under various schemes and move towards full control over management of village's natural resources.

Gram Sabha is supposed to meet once in six months. All important matters concerning village development, problems being faced by the villagers and solutions for the same are discussed in Gram Sabha. Gram Sabha elects members of the Gram Panchayat,

6.3. Gram Panchayat

Gram Panchayat is an institution of local self-government at the village level. Each Gram Panchayat consists of one or more villages depending on the population and the area. It is executive body of the Gram Sabha. The Panchayat consists of 15 to 30 members and its jurisdiction may cover population of 5,000 to 15,000. Each Panchayat is headed by an elected Sarpanch. The main function of Gram Panchayat is planning and development of the villages under its charge and addressing problems being faced by the villagers. The functions and powers of the Panchayat as envisaged are supposed to be very wide. But there are many hurdles in using those powers in actual practice.

6.4. Block Level

The block comprises about 100 villages covering population of 80,000 to 1,20,000. Panchayat Samiti or Janpada is an organ of the Panchayat Raj at the block level. It consists of all Sarpanchas of the village panchayats in the block; MLAs and MPs in the area; representatives of women, scheduled castes, scheduled tribes and cooperative societies. The Block Development Officer who is drawn from the administration functions as ex-officio secretary of the Panchayat Samiti. The main function of the Panchayat Samiti is execution of the development work in the block. The funds sanctioned by the State for development in the block are routed through Panchayat Samiti. The Block Development Office also provides technical support and guidance to the village panchayats for development works in their respective areas. It is a matter of great importance to rural people that the extension officers under his charge provide guidance to farmers in improving productivity in agriculture.

6.5. District Level

Zilla Parishad is the apex level institution of the Panchayat Raj in the District. The members of the ZP consists of all presidents of the Panchayat Samities in the district; MPs, MLAs in the district; representatives of women, scheduled castes, scheduled tribes and 2 eminent persons with experience in rural development. The number of members may vary from 40 to 70. Chief Executive Officer who is senior level officer in the administration is secretary of the ZP. The function of ZP is planning and coordination of development works in the district.

6.6. Functioning of Panchayat Raj

In spite of formal empowerment, the institution of PR is facing serious difficulties in implementing the schemes, mainly on account of inadequate as well as delay in timely availability of resources. The bureaucratic approach adopted by various state functionaries, the nodal Ministry and its departments is the main cause of the delay in providing necessary support to the panchayat raj institutions.

4. RURAL RESOURCES FOR DEVELOPMENT

1. Land

Of all the natural resources, land resource is the most important for any country. Similarly land is also the biggest rural resource. The land can be classified on the basis of topography or on the basis of land use. The topographical classification includes mountains, hills, plateaus and plains. The land use classification includes forest area, area not available for cultivation (residential, pastures, fallow etc.) and the net area sown. As per the Statistical Abstract 2003 (GOI), out of total geographical area of 328.73 million hectares (mha), reported area is 306.06 mha. Of the reported area, forest area accounts for 22.55 per cent, net area sown 46.14 per cent (out of gross cropped area of 61.99%) and the remaining 31.31 per cent includes area not available for cultivation, uncultivated and fallow land.

In the context of India, due to ever increasing population, there is tremendous pressure on the use of land. The land use pattern is closely interlinked with socio-economic factors. In past, the land excluding forest area was mainly used for residential and agricultural purposes. However during later half of 20th century and onwards, substantial growth in industrial sector has taken place, which has resulted in further pressure on the land. To accommodate some of the development projects, even some forest area had also to be surrendered. In this context, pattern of land use has to be closely related to sustainable development. Whatever may be the economic pressures in today's scenario, the importance of maintaining ecological balance in nature cannot be underestimated.

One of the ways of using the land wisely is planning future expansion of towns and cities in uncultivable land.

2. Water

Water is another resource which falls in the most important natural resources. It is required for sustenance of any kind of life on earth. In rural context, water holds the key for development of agriculture. There is no certainty about the adequate rainfall. About 70 per cent of cultivable land falls in the region of medium to low rain fall. Further monsoon period is limited to 3 to 4 months and even during this period the rains may not follow a pattern of both quantum as well as timing, which is required for the cultivation. Therefore alternative way of getting the water for irrigation assumes importance.

As there is a limit for bringing land under cultivation, only way to increase the yield from the agriculture is either by enhancing the productivity or by multiple cropping. This cannot be done without irrigation. Traditionally open wells were used for this purpose but the usefulness of open wells very much depends upon the water table which varies from season to season. Therefore agriculture has to be assisted either by providing tube wells or by irrigation projects.

These days construction of large size projects presents a lot of difficulties like large investment required, long gestation period and more importantly resettlement of project affected people. Minor irrigation is therefore a most preferred option. These include construction of bunds, ponds, water tanks and small dams & canals. Watershed management has thus acquired a lot of importance.

Due to growth of population, increasing use of irrigation for agriculture and industrial development, the use of ground water has substantially increased, resulting in lowering the water table everywhere so much so that there is scarcity of water even for drinking purpose. This calls for urgent steps for most economic use of water.

3. Forest

From ancient times forests and rivers have been considered as advancing the progress of civilization. They are the most natural and renewable resources. They prevent erosion of soil and promote rainfall. They are essential for maintenance of the ecological balance. It is a wrong notion that the forests are becoming hindrance to the industrial growth. On the contrary, they provide raw materials for the industries and contribute substantially to economic development of the country. The major raw materials include rubber, fibre, gum, timber, charcoal, bamboos etc.

The National Forest Policy was first mooted in 1894. It was revised in 1952. The main thrust of these policies is to develop the productive forests. It is estimated that the forest area of 25 to 30 per cent of the total area is a normal requirement for maintaining the ecological balance. Presently in India, the forest area constitutes 22.55 per cent of the total reported area and not more than 40 to 50 per cent of this has a good forest cover. It is observed that the environmental problems are cropping up due large-scale deforestation. The National Forest Policy (1988) therefore envisages a minimum of one-third of the total area of the country should be brought under forest or tree cover.

Tribal population depends upon forests for their livelihood and non-tribal population make use of many of the forest products such as timber, gums etc. It is therefore necessary not only to protect existing forest cover but also increase it by creating new forests or encouraging social forestry, a term used for using agricultural or other land for planting trees.

3.1. Social Forestry

Social forestry means raising of trees, grasses and fodder in agriculturists' own land, wasteland and degraded forests near to the villages. According to National Commission on Agriculture (1976) the objectives of the social forestry are to meet the requirement of fuelwood, small timber, fodder and protection of agricultural fields against winds. It also provides employment opportunity to rural people. Social forestry however needs guidance which can be best given by the extension services. Following are some of the types of forestry.

Extension Forestry: Extension Forestry covers planting of trees on wasteland and panchayat land; plantation of trees of quick growing species on lands on the sides of roads and railway lines.

A national wasteland Development Board has been established to promote planting of trees on the wasteland so as to minimize the impact of deforestation.

Agro/Farm Forestry: Agro forestry means planting trees on the bunds and boundaries of the fields. Such forestry prevents soil erosion by wind as well as floodwater during rainy season apart from economic advantages like availability of fuelwood small timber etc.

Forestry in Degraded Forests: Wherever there are degraded forests in the vicinity of the villages, such areas can be used for planting trees. Such forestry can then be useful for supply of fuelwood and small timber etc. to the villagers.

Recreational Forestry: Picnic spots can be created in the social forestry developed near the urban or rural centres, where urban people or tourists can visit the spot for recreation. Apart from getting an opportunity of seeing the beauty of nature, this can help in inculcating a feeling of friendliness towards environment.

National Thermal Power Corporation's Experiment in Developing Social Forestry: Most of the projects of NTPC are located in remote areas. Coal based thermal power stations generate large amount of ash which is dumped in ash ponds. Once the ponds are filled, trees are grown over there. This prevents not only flying of ash from the ponds but contributes to improvement in the environment. One such pond has been developed near Ramagundam in Andhra Pradesh. Eucalyptus trees are grown in the pond and it has now become a picnic spot. Interestingly this experiment has demonstrated that the trees can be grown over the ash also.

There are situations when it is unavoidable to use the forestland for setting up industries. In such situations the industries are asked to raise the social forestry of equal number of trees on the area specified by the Govt.

4. Livestock

Livestock consisting of cattle and buffalos is an important resource in rural area. India accounts for one sixth of the world's cattle and half of the world's buffalo population The livestock is extensively used in various agricultural operations in India as there is not much mechanised way of operations adopted so far. They provide valuable milk and milk products which contribute substantially to village economy. Further livestock maintenance involves lot of labour and thus it provides employment to families at home. In this context the National Commission on Agriculture specifically highlighted the role of female labour in various activities of cattle and buffalo rearing.

The average yield from Indian cattle is, however, much below the yield in developed countries. Therefore there is a need to improve the quality of livestock which can be best done by bringing improvements in breeding, maintaining hygiene and better feeding.

5. Bio-energy

The National Project on Biogas Development was started in 1981-82 with the objective of promoting family size biogas plants for providing a clean and cheap source of energy in rural area. Since then there has been a major achievement of establishing of around 3.5 million family size biogas plants and 35 million improved chulhas. The dung generated by the livestock is a great source for producing bio-gas which can be used in villages for cooking purpose thus saving of over 16 million tones of fuelwood which is brought from the forest, many a times clandestinely. The dung, organic manure, is conserved from the biogas plants and used to supplement chemical fertilizers. The organic manure amounts to 9.4 million tonnes every year. The introduction of biogas plants has resulted in improvement of sanitation, hygiene and reducing drudgery for women in kitchen work.

There are however many bio-energy plants which have been installed initially with great enthusiasm but the same are not functioning. It is necessary on the one hand to bring general awareness in rural families to keep the plants in constant use and on the other to provide assistance if there happens to be technical faults in working of the plants.

6. Human Resource

It is said that India's vast population, if their potentialities are utilised properly, can turn out to be the great strength for our country. In this context, 72.22 per cent of the total population and 58.2 per cent of the country's total work force that is engaged in the agriculture can also turn out to be the big resource for the country if they are trained properly. Excluding 35.7 per cent of the population within the age group of 0-14 years and 8 per cent of the population of the age of 60 years and above, rest of the rural population i.e. 56.3 per cent of the population is actually available for work. Further, active employment in agriculture is seasonal. It is therefore necessary to provide Extension education to agriculturists so as to enhance productivity of the agriculture. Krishi Vigyan Kendras can play important role in this. Also it is necessary to provide training facilities to rural artisans so as to enhance their potentialities to be engaged in the rural industries. Vocational training and Industrial Training Institutes can perform this function of providing training to them.

The surplus labour from rural area serves as an important input to the urban industries.

5. AGRICULTURE AND GREEN REVOLUTION IN INDIA

1. Agriculture

Agriculture constitutes about 55 per cent of the total land. It contributes about 25 per cent of GDP and about 65-70 per cent of the country's population is dependent on agriculture for their livelihood. Also it employs 58.2 per cent of the country's total workforce. Agriculture provides to the society basic necessities of life like food, fibre and raw materials. With a population of 1028.7 million to be fed, agriculture is continuing to be the mainstay of country's basic necessity as well as development. There was a time when food production in India was less than its requirement and it was dependent on other countries for meeting the gap. However after the Green Revolution has taken place in India, it is in a position not only to meet its domestic requirement but also to some extent export the food. This has happened on account of the emphasis laid on agriculture in Five Year Plans after attaining the Independence and the hard work of rural folk. Table 3.1 shows the status of Agriculture in World arena (1999). India's share in World's agriculture is quite significant and most of the varieties of crops are grown here.

Table 3.1: India in World Agriculture — 1999

Sl. No.	*Item*	*India*	*World*	*Percentage of India's Share*
I.	Area and Population			
	(a) Total Population (millions)	976	6,000	16.3
	(b) Total Area (million hectares)	329	13,387	2.5
	(c) Land Area (million hectares)	297	13,048	2.3
	(d) Arable land (million hectares)	162	1,381	11.7
	(e) Irrigated area (million hectares)	59	271	21.8
II.	Crop Production (Million Tonnes)			
	(a) Rice (Paddy)	131	596	22.0
	(b) Wheat	71	584	12.2
	(c) Cereals	230	2,064	11.1
	(d) Potatoes	23	294	7.8
	(e) Total pulses	16	59	27.1
	(f) Groundnuts (in Shell)	7	33	21.2
	(g) Rapeseed (including mustard seed)	6	43	14.0
	(h) Coffee (green)	0.27	6.48	4.2
	(i) Sugarcane	282	1275	22.1
	(j) Tea	0.75	2.87	26.1
	(k) Jute and allied fibers	2.09	3.33	62.8
	(1) Cotton (lint)	2.07	18.24	11.3
III.	Live stock (million heads)			
	(a) Cattle	215	1388	16.1
	(b) Buffaloes	92	159	57.9
IV.	Total milk (million tonnes)	77.18	561.18	13.8
V.	Tractors-in-use (thousands)	1550	26254	5.9

Source: NIRD Rural Development Statistics, 2002-03.

Although India comprises 2.5 per cent of the area of the world, India's population is 16.3 per cent (16.87% in 2001) of the world's population.

India is thus faced with a problem of increased population pressure on land and agriculture.

Land Use Pattern in India (1999-2000) is shown in Table 3.2.

Table 3.2: Land Use Pattern 1999-2000

Sl. No.	*Item*	*1999-2000 (Million Hectares)*	*Percentage to Reporting Area*
1.	Geographical area	328.7.3	
2.	Reporting area for land utilisation Statistics	306.06	
	(i) Forests	69.02	22.55
	(ii) Not available for cultivation (A+B)	42.41	13.86
	(A) Area under non-agricultural uses	22.01	7.20
	(B) Barren and unculturable land	20.40	6.66
	(iii) Other uncultivated land excluding fallow (A+B+C)	28.49	9.31
	(A) Permanent pastures and other grazing lands	11.04	3.61
	(B) Land under miscellaneous tree crops	3.62	1.18
	(C) Culturable waste land	13.83	4.52
	(iv) Fallow lands (A + B)	24.91	8.14
	(A) Fallow land other than current fallows	10.11	3.30
	(B) Current fallows	14.80	4.84
	(v) Net area sown	141.23	46.14
3.	Gross cropped area	189.74	61.99
4.	Area sown more than once	48.51	15.85
5.	Net area irrigated	56.76	18.55
6.	Gross area irrigated	77.99	25.48

Source: Statistical Abstract, India 2003 (GOI Publication)

The net area sown is 46.14 per cent. This area can be increased by about 5 per cent more by way of bringing cultivable revenue land lying idle, under cultivation.

Classification of Soil: The yield from the land depends upon the quality of soil. It is advisable to get the quality of soil tested from the soil testing laboratories. These laboratories are available in Agriculture Colleges and Agriculture Universities. If the soil is tested it can be known which particular element is deficient in the soil. This information facilitates proper selection of fertilizer as well as its correct dosage. The Tables 3.3 and 3.4 give the classification of soil based on nutrient content and mineral content in the soil.

Table 3.3: Classification of Soil Based on Nutrient Content

Classification of soil	*Organic Content (%)*	*Availability of Elements (kg/ha)*		
		Nitrogen (Natrd)	*Phosphorus (Sphurad)*	*Potassium (Palash)*
Very low	<0.20	<140	<15	<120
Low	0.21-0.40	141-280	16-30	121-180
Average	0.41-0.60	281-420	31-50	181-240
Slightly high	0.61-0.80	421-560	51-65	241-300
High	0.81-1.00	561-700	66-80	301-360
Very high	>1.00	>700	>80	>360

Source: Krishi Margdarshika 1999, Dr. Punjabrao Deshmukh Agriculture University, Akola, Maharashtra.

Table 3.4: Classification of Soil Based on Mineral Content

Classification of soil	*Availability of elements (ppm)*			
	Zinc	*Iron*	*Manganese*	*Copper*
Low	<0.6	<4.5	<1.0	<0.2
Average	0.6-1.2	4.5	1.0-2.0	0.2
High	>1.2	>4.5	>2.0	>0.2

Source: Krishi Margdarshika 1999, Dr. Punjabrao Deshmukh Agriculture University, Akola, Maharashtra.

2. Intensive Agricultural District Programme (IADP)/ Intensive Agricultural Area Programme (IAAP)

Intensive Agricultural District Programme was started in 1960-61. The programme was initiated as a result of the recommendations of the team of experts from Ford Foundation, USA in 1959 regarding special steps to be taken for improvement of productivity of agriculture. During 1960s there was limitation on providing required inputs like improved quality of seeds, fertilizers, extension education and extension staff all over the country. It was thus not practicable to take up intensive agricultural programme on the large scale. The programme was therefore initiated in seven districts selected on the basis of availability of irrigation facility, institutional set up like co-operative societies, panchayats and districts not affected by floods, droughts etc.

The Central as well as State Govt. supported the programme by way of providing administrative and extension staff, strengthening co-operative institutions, adequate and timely supply of credit and all physical inputs like seeds, fertilizers, pesticides and implements through co-operatives. As a result of the concentrated effort, there was significant improvement in the productivity of the agricultural produce.

In view of the encouraging results of IADP, the programme namely Intensive Agricultural Area Programme (IAAP) was taken up in 1964-65 on the similar lines to cover wider area. Package Approach for providing all necessary inputs and facilities was adopted in the programme and it covered development of important crops like cotton, pulses, wheat, paddy etc. in 114 districts. It was assumed that the development in the selected areas would set the pace of development in the adjoining areas.

However the above intensive programmes had certain shortcomings. The focus of these programmes was not on the development of irrigation facility. The programmes envisaged use of improved seeds and high yielding variety (HYV) seeds were yet to be developed then. Further the programmes were based on concentration (on certain selected areas) principle rather than country as a whole. With passage of time, Indian agriculture in later period witnessed number of developments. The schemes to tap the irrigation potential were started. The availability of HYV seeds and fertilizers in the market improved. The position of availability of credit also showed improvement. In view of these new developments, the above programmes gave way to general approach towards development of agriculture all over the country.

3. Irrigation

Agriculture in India is beset with number of problems. Nearly 65 per cent of the cropped area is dependent on the rain and as such when the monsoon is erratic, there is a problem in the output. In this context irrigation of any kind becomes important for the increasing the productivity of agriculture. It is estimated that the productivity in irrigated farming is 2.5 tonnes per hectare as against 1 tonne in dry land farming. Irrigation area is categorised in two types. The net irrigated area is the area irrigated during an agricultural year (July-June) counting the area only once even if two or more crops are irrigated in different seasons on the same land. The gross irrigated area is the total irrigated area under various crops during the year, counting the area irrigated under more than one crop during the same year as many times as the number of crops grown, crops sown mixed being taken as one crop. From this it follows that the area irrigated more than once is obtained by deducting the net area irrigated from the gross area irrigated. Similarly area sown more than once is obtained by deducting net area sown from the gross cropped area. Taking the various factors into account, the ultimate irrigation potential in India is estimated to be around 139.88 mha and as on 31.03.2002, 93.98 mha has been created with major and medium projects contributing 37.08 mha and minor irrigation contributing 56.90 mha.

In 1999-2000, out of 141.23 mha of net area sown, net area irrigated is 56.76 mha i.e. 40.12 per cent. In 2000-01, net area irrigated is less i.e. 54.68. Assuming net area sown as the same as 141.23 mha, net area irrigated is 38.72 per cent of net area sown. These figures also include the irrigation under the conditions of rainfall and if such land is excluded, the area irrigated will be about 35 per cent of the cropped area.

It is disturbing to note that the irrigated area has decreased in 2000-2001 compaired to 1999-2000 as given below:

Year	*Net area irrigation (mha)*	*Gross area irrigated (mha)*
1999-2000	56.76	77.99
2000-2001	54.68	75.14

3.1 Classification of Irrigation Works

The classification of irrigation works is based upon cultivable command area (CCA). Projects using surface waters each with a CCA of 10,000 hectares and above are classified as major irrigation projects, with CCA between 2,000 and 10,000 as medium irrigation projects and with CCA below 2,000 as minor irrigation projects.

Minor irrigation through open wells, tube wells, water tanks etc. constitutes major portion (65.48 per cent) of the net irrigated area. Table 3.5 shows the break up of different sources of irrigation which aggregate to 54.68 mha of irrigated area in 2000-01.

Table 3.5: Net Area Under Irrigation by Sources 2000-01

Source		*Net area under irrigation ('000 hectare)*	*(Per cent)*
Canals	Govt.	15789	28.87
	Private	199	0.36
Tanks		2524	4.62
Tube Wells and other Wells		33277	60.86
Other sources		2892	5.29
Total		54682	100

Source: Statistical Abstract 2003 (GOI Publication)

However the stage has now come, at some places water table has gone down to such a low level that it is difficult to have it as a perennial source. Notwithstanding this, there is a potential of irrigation elsewhere, although of varying degree, and the same needs to be tapped. This calls for better watershed management, which essentially involves capturing water in the tanks during rainy season and then using it judiciously during remaining part of the year. In essence the rainwater flowing through rivers and *nallahs* has to be harvested either by way of constructing dams or watershed management.

The small and marginal holdings constitute 80 per cent of the total land holdings. The land holdings are steadily getting fragmented on account of divisions in the rural families from generation to generation and small and marginal farmers are finding quite difficult to meet both ends. However, application of technological advances, provision of effective extension services, availability of inputs (within reach) like fertilizers, pesticides, credit etc are likely to improve the overall agricultural scenario. Rural population also comprises of a large number of landless people who are below poverty line. There is also undue pressure on land which is not available in unlimited measure.

Therefore rural industrialisation, provision of institutional credit with supporting services for permanent income generating schemes, payment of minimum wages to agricultural labour, providing social security measures for rural poor, etc. call for greater attention of the Govt.

4. Green Revolution

By 1960-61, the stage has reached in India when the productivity with the conventionai mode of agriculture reached a plateau. Thereafter efforts of research institutions, both national and international, in the field of agriculture were directed to improve the quality of seeds as well as fertilizers. These efforts led to the "Seed-Fertilizer Revolution" with the production of high yielding variety (HYV) of seeds which together with fertilizers started giving a much higher yield. The spectacular increase in foodgrain production between 1966 and 1971 marks the beginning of the Green Revolution in India. Since then India has come a long way in achieving higher production in all the foodgrain crops. Table 3.6 shows the production of principal crops and items allied to agriculture from 1950-51 to 2002-03.

Table 3.6: Production of Principal Crops and Items Allied to Agriculture (1950-51 to 2002-03)

(Million tonnes)

Crops	*1950-51*	*1970-71*	*1980-81*	*1990-91*	*2001-02*	*2002-03*	*2003-04*	*2004-05*
Total Food grains	50.82	108.42	129.59	179.39	212.85	174.19	212.0	206.4
Rice	20.58	42.22	53.63	74.29	93.34	72.65	87.0	87.8
Wheat	6.46	23.83	36.31	55.14	72.77	65.10	72.1	73.0
Other Cereals	15.38	30.55	29.02	32.70	33.37	25.30	37.7	31.9
Pulses	8.41	11.82	10.63	14.26	13.37	11.14	15.2	13.7
Oilseeds	5.16	9.63	9.37	18.61	20.66	15.06	25.1	24.8
Sugarcane	57.05	126.37	154.25	241.05	297.21	271.58		
Cotton (bales)	3.04	4.76	7.01	9.84	10.00	8.72		
Milk	17.00	21.20	31.60	53.90	84.75	NA		
Fish	0.80	1.80	2.40	3.80	5.94	6.20		

Source: Statistical abstract, India 2003, GOI Publication & Economic Survey 2004-05, GoI.

The use of HYV seeds led to many changes in the agricultural inputs. Irrigation is the prerequisite for use of these seeds. Thus, more stress had to be laid on taking up minor and major irrigation projects. The HYV seeds are highly responsive to fertilizers, as such the demand for fertilizers has increased. The HYV crop is susceptible to diseases and pests, as such the demand for the pesticides and sprayers has also increased. The higher production called for availability of marketing and credit facility on the higher scale. The Green Revolution in effect changed the whole set of practices which were being followed since ages. It resulted not only in higher productivity to farmers but with its forward linkages generated high growth for industries to produce large quantities of pumpsets, fertilizers, pesticides, sprayers, threshers etc. and also generated more demand on electricity. However, along with increase in productivity, the input costs to agriculture have gone appreciably high.

Earlier there was a time when India had to take recourse to import the foodgrains to bridge the gap between the country's requirement and indigenous production. The effect of the Green Revolution as well as improved contribution of non-irrigated farming is seen in, India becoming not only self sufficient in food but also exporting the produce. Fig. 3.1 shows agricultural production, cropped area and land utilization from 1990-91 to 2003-04.

Index No. of Agricultural production
(Base: Crops Years Triennium Ending 1981-82 = 100)

ALL CROPS

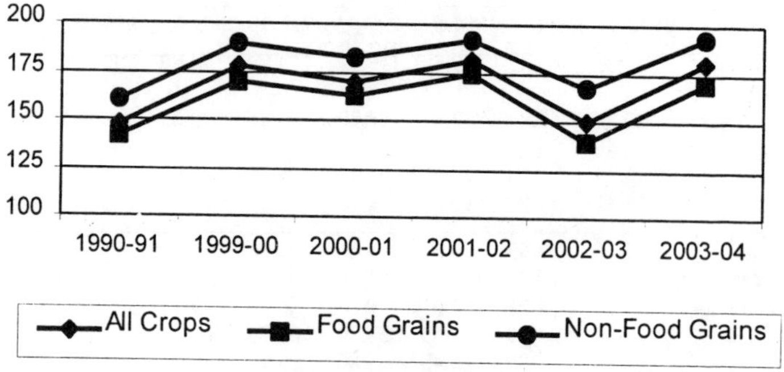

CROPPED AREA

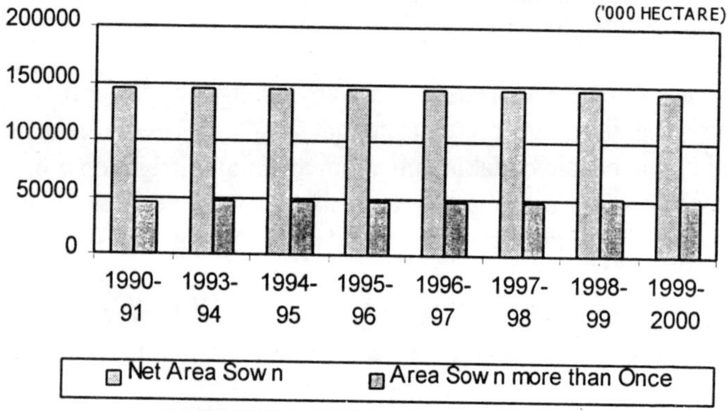

LAND UTILISATION (1999-000)

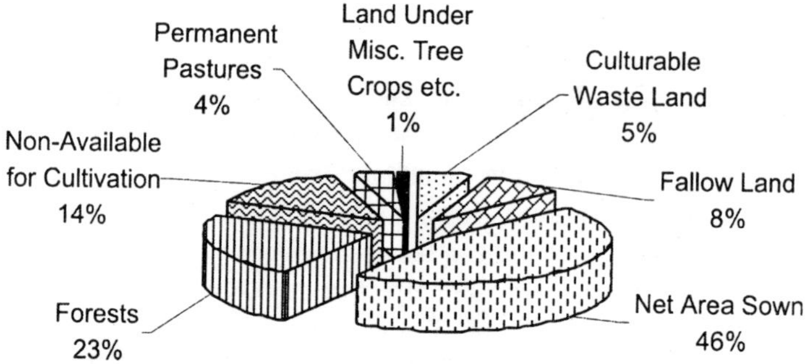

Source: Statistical Abstract, India 2003, GOI Publication.

Fig. 3.1: Agricultural Production, Cropped Area and Land Utilisation

5. Extension Education

5.1. Meaning and Scope of Extension Education

Extension education means educating the farmers about the methods by which quality of produce as well as their productivity can be improved. It was in USA that the extension education was first started as a subject of great importance. In India the villagers are mostly uneducated and they are used to do agriculture as per the old practice. They are not generally aware of the new developments that are taking place around in various fields of agriculture. The extension education covers all aspects of agriculture like testing of soil, selection of crops suitable for the soil, rotation of crops, proper ploughing, selection of quality seeds, timely and proper sowing, use of proper fertilizers, timely removal of weeds, use of effective pesticides, timely and proper harvesting and preserving the produce before marketing. It involves procedural as well as technical aspects.

5.2. Early Efforts in Extension Education

During Pre-Independence period efforts were made by eminent persons to experiment new ideas of extension education. These were known as Sriniketan, Martandam, Gurgaon, Baroda, Firka, Etawah, Nilokheri and Sewagram projects/experiments. As already discussed, these projects were aimed at providing extension education in respect of improved agricultural practices, developing institutions like co-operatives, providing adult education and reforms in many other social and economic practices.

5.3. Organisation Structure

The extension education forms part of the responsibility of Agriculture Officer. The Extension Worker/Agriculture Supervisor who reports to Agriculture officer looks after a village or the number of villages depending upon the number of farmers. The Panchayat Raj Institutions play a vital role in supporting the extension education programme. The selection of the extension worker and training for him are important aspects. Looking to the importance of the agricultural programme, all Extension Workers/Agriculture Supervisors should be graduates in agricultural science (BSc. Agri).

5.4. Implementation of Extension Programme

The farmers are accustomed to do agriculture in a traditional way. They will not modify their age-old practice unless they are convinced about the advantages that the modification will bring to them. Further any changes to be brought about should be in harmony with the existing environment. The task of extension agency therefore starts with study and survey of the local situation as exists in the village. It includes social aspects as well as other aspects such as skill that the local people possess; mechanical implements being used; crops being generally produced; agricultural practices being followed; financial status of big, small and marginal farmers and landless labourers etc. Among various types of improvements that the research and development suggest, the extension officer has to choose the particular modifications that will suit the farmers at the local level. There are situations when explaining verbally does not suffice. In order to make it effective, he has to demonstrate the benefit in physical terms. This is best done by selecting two plots of the land, sowing in one the usual crop and adopting the usual practice and in another the new variety of crop and adopting there modified practice. If such a demonstration proves benefit in the form of improved quality of produce, better productivity and the overall cost effectiveness, the farmers will be more willing to follow the advice of extension agency.

Once the farmers start using the modified practice, it is incumbent on the part of extension agency to make necessary arrangements for maintaining supply chain of the inputs that are required in the modified process. Otherwise farmers will be discouraged in following any suggestion for modification. The point is that the modification has to be taken to the logical end so that the farmers continue to reap the benefit when they follow the advice given by the extension agency.

The extension education comprises educating the farmers about the research and development that are taking place in agricultural products and the services. When the new practices are implemented, farmers may come across various difficulties. The extension agency has therefore to give inputs to the other end i.e. research and development centres about the benefit or the practical problems encountered in implementing the modifications. This feedback will enable the R&D centres to proceed with further research or offer suitable solutions for the problems being faced in implementation of the innovations suggested by them.

5.5. Community Participation

Finally the farmers' community has to get convinced that the extension programme is for their benefit and they must get a feeling that it is their own programme. No programme will be successful if the community feels that it is being thrust on them by the administration. The approach of the programme should be such that the villagers take it as a self-help programme. The community participation can be helped by identifying the change agents from farmers and leaders who are sympathetic to the programme and seeking community participation through them.

5.6. Mode of Imparting Extension Education

The extension education can be imparted through formal courses by conducting the classes for the farmers. It can also be done through non-formal way by discussing in the village groups. In such a scheme the place does not matter as it can be done even under a tree. The education can be imparted through individual contact by way of paying visit to the farm or homes; group contacts by way of group discussion, method-demonstration, field trips; mass contact by way of publication of news magazines, bulletins, leaflets, pamphlets, organising farm fairs, special campaigns, exhibitions, audio-visual presentations, and publicity through radio and television.

5.7. Evaluation of Programmes

Periodic evaluation of the programme under implementation is very essential and this aspect forms an integral part of any extension programme. Many a times it so happens that the farmers encounter certain difficulties in implementation and they are just unable to take timely corrective action. The assessment of the programmes need not be on individual basis. If there are many programmes running simultaneously, the assessment can be done on any one of them. The evaluation should encompass all important aspects like improvement in quality, increase in productivity and differential cost i.e. the cost involved in the earlier practice and cost involved in adopting the new practice and the gain as a result of adopting the new practice. Such evaluations will enable the farmers as well as the administration to decide about the efficacy of the programmes undertaken.

6. LAND REFORMS

1. Historical Background/Need for Land Reforms

There is a distinction between the land reforms and agrarian reforms. The land reforms relate to structural changes in the size distribution of land holdings and establishment of a system of tenure. The agrarian reforms involve modification of wide range of conditions relating to the agricultural sector, which may include land reforms. The rural development is still a wider field which relate to, beside agriculture, other areas like education, healthcare etc.

There has always been pressure on land as the people who are dependent on agriculture are many but the land is limited. Earlier the ownership of land was much uneven. There were big landlords, medium landowners and landless labour. It was impossible for the big landlords to cultivate the entire land by themselves and therefore they were either leasing the land to tenants or big chunk of land used to remain untilled. The leasing of land to tenants led to lesser capital investment by the tenant and consequently resulted in lesser productivity in the agriculture. The tenants are also subjected to exploitation by the landlords by way of terminating their tenancy anytime or increasing the rent for cultivation. While some tenants used to have occupancy rights, mostly they have been cultivating the land on oral understanding only. On the other hand landless labour have been remaining idle except during the peak agricultural season of four months. It is in this context that the land reforms acquired great urgency.

As per Constitution of India, "land, that is to say, rights in or over land, land tenures including the relationship of landlord and tenant, and the collection of rents; transfer and alienation of agricultural land; land improvement and agricultural loans; colonisation" falls in the State List at sl. no. 18. Accordingly the states have passed legislations as per exigencies of the situation as existing in the states. Therefore there is no uniformity in the land laws throughout the country.

2. Approach to Implementation

To begin with it was herculean task to make the landowners mentally and physically prepared to part with the land even if it is surplus. It is in this context that Acharya Vinoba Bhave, staunch follower of Gandhiji, spearheaded a "*Bhudan*" movement in 1951 which received spontaneous support from some rich landlords in voluntarily donating part of their land. He expounded the principle that the land belongs to motherland and no individual can claim to be its sole owner. During those days this movement got very good response especially in Bihar.

In the context of larger problem of redistribution of gigantic area of land throughout the country, the voluntary approach adopted in *Bhudan* movement obviously had its own limitations. The Govt. thus moved towards implementing land reforms encompassing important areas like abolition of intermediaries, ensuring security of tenure of the tenant, determining surplus land as per land ceiling act and its redistribution.

2.1 Abolition of Intermediaries

Intermediaries mean those persons who are in between the state and the actual cultivators. There are landlords who instead of cultivating the land by themselves engage the tenants for the purpose. This happens if the land owned by the landlord is more than his tilling capacity or if the landlord is an absentee landlord or for any other reason. After India got Independence, the Govt. took up the first task of abolition of *jamindari* and *jagirdari* in order to remove the intermediaries between the state and the cultivators. The legislation known as Abolition of *Jagirdari/Jamindari* to this effect were passed between 1948-54 by almost all the states.

As per fifth Five Year Plan, the abolition of Jamindari system is estimated to have benefited about 20 million tenants by 1974-75. Some landlords however continued to hold large estates through loopholes in the system. Also the large mass of sharecroppers and landless labourers did not get benefit from the system. Notwithstanding these shortcomings, the abolition of Jamindari proved to be a good step in the direction of implementing land reforms.

2.2 Tenancy—Security of Tenure

Tenants fall into different categories. Some are permanent tenants who cannot be evicted so long as they pay the rent regularly. Some are temporary tenants who are given the land for tilling from time to time either on written or oral agreement basis. The major problems of tenancy are lack of security and hiking of rents. Laws have been enacted by the states with regard to security of tenure to prevent large- scale eviction of tenants. According to these laws, the tenants cannot be evicted unless the landowner wants it for personal cultivation and even in such a case prescribed minimum area has to be left with the tenant. The number of such protected tenants is gradually coming down.

2.3 Land Ceiling Act

The most important feature of the land reforms is to fix the upper limit on the land holding and declare the land falling above this limit as surplus. The next task of the state is to take possession of this surplus land and then distribute it to the landless or marginal/small farmers. The fixation of this ceiling limit by every state was no easy exercise. However almost all the states passed the land ceiling acts with the following important features:

2.3.1 Principle of Application of Ceiling Limit

In rural area, agriculture is the main occupation and it is a source of employment as well as income for the livelihood of the vast majority of people. Therefore land ceiling should apply to only those who have more land than their requirement.

2.3.2 Application of Ceiling to All Categories of Land

The land falls in various categories like fallow, uncultivable, cultivable and irrigated. The ceiling, though may be graded, should be applicable to all categories of land. The underlying idea is that fallow and uncultivable land can either be brought under cultivation, may be with some extra efforts, or it can be used as pastures for cattle.

2.3.3 Ceiling for Irrigated Land

The productivity of irrigated land is more. Further there is an irrigation facility by which more than one crop can be taken during the year. Based on this principle, there are two separate ceiling limits fixed for such categories. In case of land which is irrigated and yields two crops a year, the limit is 18 acres. In case of land which is irrigated and yields one crop a year, the limit is 27 acres.

2.3.4 Ceiling for Rest of the Land

For all and different kinds of lands beyond the ones mentioned at sl. no. 3 above, the limit is 54 acres.

The ceiling limits though vary from state to state, most of the states have adopted the above limits. Based on the principle of equity and social justice in land holding. State Governments passed legislations for fixing ceiling on land holding along with certain exemptions. The exemptions relate to plantation of tea, coffee etc.; land held by industrial and commercial undertakings and surgar factories. The legislations were enacted by 1961-62. Land reforms in larger context include protection of tenants from increasing the rent and eviction by the landlords. It encourages sale of land, if any, to the occupying tenants. A need has also been felt to provide special protection to tribals and scheduled castes against grabbing of their land by other landlords on some pretext or the other.

It is pertinent to note that the tenants as well as landless persons to whom surplus land is distributed are not by themselves self supporting and therefore they need institutional help in the form of extension education and credit etc. for arranging inputs for the agriculture so as to make the land given to them viable.

Land holding is getting fragmented from generation to generation due to divisions in the joint families. Against this background, consolidation of holdings which means pooling of land of a group of villagers or of the entire village and then cultivating it on cooperative principle is also being tried in various places with varying degree of success. To maintain and update huge land records, in a systematic way, following Centrally sponsored schemes have been started:

2.4 Computerisation of Land Records (CLR)

The Centrally sponsored Computerisation of Land Records Scheme was started in 1988-89 as a pilot project in eight districts, one in each state with a view to remove the problems inherent in the manual systems of maintenance and updating of land records and to meet the requirements of the various groups of users. At present the scheme has been started in all districts of the country.

2.5 Strengthening of Revenue Administration and Updating of Land Records (SRA&ULR)

The Centrally sponsored Strengthening of Revenue Administration and Updating of Land Records Scheme was started in 1987 with a view to assist the States in the task of updating the land records. Under the scheme, financial assistance is given for purchasing modem survey equipments like Global Positioning System (GPS), EDM, Total Stations, Theodolites, Work Stations, office equipments etc. to take up the fieldwork. However the actual progress of implementation of both the above schemes at ground level is quite slow.

3. Shortcomings of Land Reforms

The land reform process encompasses two distinct aspects, one relating to ownership and the other related to operational aspect that is tilling. While there has been definite improvement on both the fronts after the land ceiling acts were introduced by various states, the objectives of land reform could not be fully realised. The main shortcomings in the implementation of the reforms have been discussed below.

3.1 Nature of Surplus Land

There would always be a tendency on the part of any landowner to retain within the ceiling the best part of the land and part with the rest In this process it so happened that the land declared as surplus beyond the ceiling thus comprised of such a land which is largely fallow or difficult to cultivate. The question arises how poor peasants can afford to make such a land cultivable unless the Govt. provides the cultivators required technical (in terms of extension service) and financial support.

3.2 Inequalities in Ownership

As per the Annual Report 2002-2003 of Ministry of Rural Development (GOI), by 31st March 2002, out of 7.37 million acres declared as surplus, possession of 6.5 million acres was taken over and out of that 5.39 million acres were actually distributed. It is also observed that the pattern of land ownership, even after more than 42 years of enactment of land ceiling act, is skewed in the sense that a very few own fairly large extent of land and large number have very small size holdings or are landless. 59 per cent of the total holdings constitute as marginal holdings with less than one hectare. 19 per cent of the total holdings constitutes as small holdings with one to two hectares of land. Thus small and marginal holdings together constitute about 80 per cent of the total Holdings.

3.3 Lack of Political Will

Any reform process is a difficult one as it is a change from the situation as existed for a long time. The process of redistribution of land is especially more challenging as it involves transfer of ownership of supposedly most valuable permanent asset and a system that existed for centuries. Such a reform therefore calls for high degree of political will. This aspect is evident from the varying degree of success of implementing the land reforms in different states. For example there has been less impact of the land reforms in states like Bihar, Madhya Pradesh, Uttar Pradesh, Andhra Pradesh, Orissa, Gujarat and Haryana. On the other hand the impact of reforms in West Bengal has been striking. Political will and sincerity of bureaucracy make a lot of difference in dealing with such crucial matters.

7. RURAL MARKETING

1. Market Systems

This is a very crucial area for the farmer. After undergoing grueling exercise of completing all agricultural activities, when the produce is ready to be sold, farmer obviously looks forward to price friendly market. Having spent money whatever he had on cultivation, farmer is in dire need of prompt payment for his produce for meeting family expenditure as well as repayment of any loans taken earlier. Except for rich farmers, small and marginal farmers are not in a position to wait any longer for the return. Setting aside their own requirement, these farmers therefore proceed for selling their marketable surplus. At this point of time when large quantity of marketable new produce arrives in the market, prices are generally depressed. As small and marginal farmers can not afford to wait inordinately for the payment, middlemen taking advantage of the situation purchase it at lower prices and then keep it in storage sheds so that the same can be sold at higher price at a later date.

Agricultural Produce Marketing Committee (APMC), a cooperative body of agriculturists, is the best marketing mechanism intended to serve the farmers. However to take advantage of this system, farmers face twofold difficulties. One is that the APMCs' storage facilities are generally located at district and taluk places and it becomes quite difficult for individual farmer to arrange transportation to these far off places. The other difficulty is getting immediate payment. APMC have two systems. According to one system, farmer can use its storage facility by paying rental charges towards storage and he can sell his produce at a time of his choosing. In this case farmer keeps ownership with him. The other system is farmer surrenders ownership of his produce to APMC and APMC sells the produce in open auction which is held when the prices are favourable. The farmer gets a loan to the extent of 60 per cent of the ruling price at the time of handing over the produce to the Committee for which he has to pay the interest charges which are of course generally lower than the bank rates. After the auction when the Committee receives the payment from the bulk purchasers, it makes final payment to the farmer after adjusting the advance. Some improvements are needed in the marketing system such as APMC should open their branches at sub centres so that it is within a reasonable distance of 3 to 5 km from any village. Also the amount of advance as a percentage of the market price given to the farmer can be increased based upon past market price behaviour.

2. Grading and Weighment

The basic purpose of grading is to encourage the farmer to produce a quality commodity and for which he should get appropriate price. If applied properly, this goes in favour of the farmer as well as consumer. However it is unfortunate that the way the grading system is implemented at the ground level, average farmer is generally a looser. This happens because of the attitude of some unscrupulous grading staff of APMC. The subjectivity in adjudging the quality is fully utilised to favour some farmers and downgrade the others. Similarly manipulation also takes place in weighment of the commodity.

In the above context, the Maharashtra state sponsored cotton procurement scheme is the glaring example. The scheme could not succeed mainly on account of mismanagement of grading and weighment system. Besides, the State has incurred cumulative losses due to more price paid to farmers than the market price. This has been done with a view of giving remunerative price to the farmers.

If the above aspects are properly taken care of, APMC and the State sponsored procurement schemes can be the best options for the farmers, thus avoiding their exploitation by the middlemen and the speculative market forces.

8. RURAL CREDIT FACILITIES

1 Nature of Facilities Required

8.1. Specific Needs of Rural Credit

Majority of the rural population depends upon agriculture and as such they have peculiar requirement of credit as compared to urban population or industrial sector. Rural sector has many diversities and it is like an unorganised sector. When we envisage rural credit, we relate it for the needs of rural poor. Requirement of credit in rural area arise for payment of old debts like inherited debt or for current needs either for needs of the family or for doing the agricultural operations. The requirements are: (i) short term i.e. for purchase of seeds, fertilizers, pesticides etc, (ii) medium term i.e. for construction of well for the water source or purchase of implements, (iii) long term i.e. for purchase of land, tractor-trailer etc.

8.2. Expectations of Rural Credit Seekers

The credit need in rural area is overwhelmingly agriculture oriented and therefore it has to be available in time for doing particular agricultural operation and adequate enough to be useful. Majority of the rural population is not well educated and they are not conversant with the nitty gritty of the formalities to be completed for getting the loan. Further they cannot afford to be away from their work places often. Therefore the whole system has to be simplified in such a way that there are no hassles for the agriculturists. Also from the point of view of both creditor as well as borrower, there is a need to amalgamate credit with extension education so that the credit is effectively used for the purpose for which it is given.

2. Earlier Sources of Credit

Before institutional sources came into existence as we witness today, rural people used to depend upon local moneylenders popularly known as *Sahukars*. The system entailed many disadvantages like: (i) high interest rates were charged, (ii) when the output of agricultural produce was less during the year, only interest could be paid, (iii) many a times even the payment of interest was not recorded in the loan paper. The system thus led to borrower becoming perpetual debtor. At times when the farmers were unable to repay the loan, either the produce used to be mortgaged to the *Sahukar* or sometimes even land used to be taken over by the *Sahukar*. However this system also had an advantage, that is, borrower used to get the loan instantly and did not involve elaborate paperwork. It also depended upon the nature of the *Sahukar* in the sense that a well-meaning *Sahukar* having personal knowledge of the conditions of the borrower could take sympathetic view about the repayment. There existed a personal chord between the two and the lender usually participated in various family functions of the borrower such as marriage ceremony etc. However with a few exceptions, the system as a whole was beset with many evils.

Experiencing the evil of exploitation by the moneylenders and taking into account increasing quantum of credit requirement as a prelude to adaptation of farm technology, it became necessary to establish institutional credit framework in the rural area. This led to establishment of cooperatives and commercial banks.

3. Co-Operatives

3.1. Meaning of Co-operation

The meaning of cooperation is working or acting together and forming a co-operative is institutionalising the concept of cooperation. A need for cooperative has been felt by the not so rich people because they are unable to conduct their work individually. The concept of co-operative grew initially with the idea of obtaining the credit. It has a special relevance to the agricultural sector in India as 80 per cent of the farmers are small and marginal. These farmers face financial difficulties at every stage of agricultural operations i.e. preparation of land; buying seeds, fertilizers, pesticides; sowing; harvesting; storing the produce; transporting and marketing the produce. As such their need for institutional credit is well recognised.

3.2. Principles of Co-operation/Co-operative

The co-operatives are based upon following basic principles.

(i) Members having interest of common welfare come together to form the co-operative.
(ii) It is purely a voluntary organisation.
(iii) All members will be treated as equal.
(iv) The members will choose the office bearers of the organisation for conducting its day-to-day business.
(v) It presupposes that the office bearers of the organisation will be fired with the fervour of public service rather than extracting benefit for themselves.
(vi) The co-operative will work towards common goal rather than for good of any particular individual or a specific group of people.
(vii) Members will have a full say in formulation of the policies of the co-operative.
(viii) The working of the co-operative will be fully democratic.
(ix) The co-operative will work as economically sustainable organisation and will strive to earn reasonable return on investment of the members.

4. Historical Background

The condition of agriculturists during pre-Independence era was pretty bad. Credit was always a serious problem for them. It was in 1895 that the problem of rural credit came to be highlighted through a report of Frederick Nicholson. Famine Commission 1901 too highlighted the credit problem. These reports culminated in passing of the first Co-operative Credit Societies Act in 1904. The salient features of the act were, (i) Any ten persons in the village or town can form the co-operative society, (ii) society can raise the funds by way of obtaining deposits from the members as well as from other persons, Govt. or other co-operative credit societies, (iii) society can offer loans to members or to other co-operative credit societies with the permission of Registrar of the co-operative societies.

The loans were sanctioned by the primary co-operative societies on the basis of immovable assets and not movable assets except ornaments. The principles of Raiffaison and Schulze who were pioneers of co-operation movement in Germany were generally treated as guiding principles in organising the societies in India.

The 1904 Act however suffered from many shortcomings, major one being there was no supporting financing mechanism, in terms of central bank, either specified or in place. These shortcomings were removed in the Act of 1912. The new Act further widened the scope of Co-operatives by way of envisaging the formation of co-operatives for farming, marketing, housing, commodity distribution etc. It also envisaged formation of central and provincial banks.

In 1914, Govt. appointed Maclagan Committee to review the working of the 1912 Co-operative Act. The Committee opined that the co-operative societies are not functioning in the true spirit of co-operation and suggested certain improvements such as following proper procedure of accounting and audit and training of members and office bearers. The later period saw formation of a spate of co-operative societies.

After Second World War, recognising the important role of co-operative movement, the Govt. appointed Co-operative Planning Committee in 1945. This was followed by number of committees i.e. Gadgil Committee 1945, Saraiya Committee 1946 and Manilal Nanavati Committee 1947 to examine the working of the co-operatives and suggest measures for improvements. The Saraiya Committee recommended that the primary credit co-operative societies should be made multipurpose, the Govt. should contribute towards the share capital of central and apex co-operative banks and the Govt. should be associated with the managements of the co-operative societies so as to ensure interests of the worthy agriculturists. The Govt. accepted these recommendations.

In 1951, Reserve Bank of India (RBI) appointed a committee named All-India Rural Credit Survey. The Committee submitted its recommendations in 1954. The most important recommendations included:

(i) Establishment of National Agricultural Credit (Long-Term Operations) Fund to give loans to State Governments to subscribe to the share capital of co-operative institutions and providing medium term loans to the State Co-operative Banks

(Apex Banks) for enabling them to provide finance for agricultural and allied activities.

(ii) Establishment of National Agricultural Credit (Stabilisation) Fund to provide funds to the State Co-operative Banks for converting short-term loans due from the State Co-operative Banks to the RBI into medium-term loans in critical periods like natural calamities when it is impossible to repay the short-term loan.

(iii) Establishment of a Central Warehousing Corporation with a network of warehouses throughout the country for storage of agricultural produce

(iv) Establishment of Central Land Mortgage Bank in each state and Primary Land Mortgage Banks at the lower levels.

The above recommendations were accepted by the Government. The Govt. appointed Mehta Committee in 1959 to review the working of the co-operative societies and suggest improvements. The important recommendations of the committee included:

(i) The primary co-operative credit societies should be reorganized as viable units with village as a basic unit and each society with adjacent villages covering a population of 5,000.

(ii) The short-term and medium-term loans up to Rs. 5,000 should be given to agriculturists without the mortgage of landed property.

(iii) The State should participate in the share capital of primary agricultural cooperative credit society to the extent of Rs. 5,000 per society.

(iv) Credit limits should be liberalised in the case of Central Co-operative Banks (at the district level) to enable them to obtain more funds from RBI (NABARD) to enable them to provide higher amounts to primary societies.

Most of the recommendations of the Mehta Committee were accepted by the Government.

Venkatappaiah Review Committee Report 1969 pointed out the huge gap between the requirement of the credit and the amount of credit being made available to the agricultural sector. According to the report only 25 per cent of the total credit need was being met. Apart from increasing the credit, it also emphasised to adopt the multi-agency approach in delivery of the credit. Thus a need for enhancing the agricultural credit has been well recognised.

There has been steady increase in the investment in agriculture all these years. However the investment as a percentage of GDP is consistently going down. This is bound to create serious difficulties in agricultural sector in the long run. The Budget 2004, taking note of the greater need of the agricultural credit, intends to double the flow of agricultural credit in three years. The implementation of the policy on agricultural credit announced on 18th June, 2004 is entrusted to public sector and private sector banks, Regional Rural Banks (RRBs) and the Cooperative Banks.

5. Rural Credit Banks

5.1. National Bank for Agriculture and Rural Development (NABARD)

Earlier, the Reserve Bank of India, set up in 1935, was looking after credit needs of the agricultural sector. To meet the credit requirement of the farmers, RBI provided short and medium term credit to State Co-operative Banks. Similarly it provided long-term credit to Central Land Development Banks. However as the credit requirement increased manifold, Govt. appointed a Committee to Review Arrangement for Institutional Credit for Agriculture and Rural Development (CRAFICARD) in 1979 under the Chairmanship of Shri B. Shivaraman. Based on the recommendations of the Committee, Govt. of India set up National Bank for Agriculture and Rural Development (NABARD) in 1982, with the objective of providing full attention to the credit requirements of the rural area.

Board of Directors of the bank is appointed by the Govt. of India in consultation with Reserve Bank India. It consists of representatives of Reserve Bank of India, cooperative and commercial banking, Govt. Of India, State Governments and experts in the field of agriculture and rural development.

The authorised capital of NABARD is shared by Govt. Of India and Reserve Bank India equally. The functions of the bank includes:

(i) Planning of operational matters, coordination, monitoring, research, training, etc. with regard to rural credit;

(ii) Providing refinance assistance to the State Co-operative (Apex) Banks, Regional Rural Banks (RRBs) and any other financial institutions as recommended by Reserve Bank of India. It ensures that at least 20 per cent credit is earmarked by banks for financing small and marginal fanners;

(iii) Providing loans and advances to State Governments to enable them to subscribe to the share capital of co-operative societies;

(iv) Undertaking inspection of co-operative societies other than primary co-operative credit societies and Regional Rural Banks.

(v) Maintaining National Rural Credit (Long Term Operations) and the National Rural Credit (Stabilisation) Funds which are contributed by Central and State Governments. Based on disbursements so far it is observed that the major beneficiaries of the NABARD funds are State Cooperative Agriculture and Rural Development Banks, Commercial Banks and Regional Rural Banks followed by State Co-operative Banks.

5.2. State Cooperative Banks (SCBs)

This is apex level bank in the 3-tier system in the state for providing short and medium term loans. The State Co-operative bank is formed by federating District Central Co-operative Banks. In each State there is a State Cooperative Bank with its branches in the regions. The funds for State Co-operative Banks are secured by way of share capital subscribed by the affiliated District Central Cooperative Banks (DCCBs), deposits from the public etc. These banks get major part of the funds from NABARD and provide in turn financial assistance to District Central Cooperative Banks.

There are 29 (March 2001) State Co-operative Banks in the country. The main source of working capital for these bank is Reserve Bank of India. The performance of the banks vary from State to State. The performance of State Co-operative Banks in Maharashtra and Gujarat has been much better than in other states. The co-operative movement in these states has always been in the forefront. However lately there have been instances of gross mismanagement in some of the cooperative banks.

5.3. District Central Cooperative Bank (DCCB)

This is at the middle level of the 3-tier system for providing short and medium term loans. The District Central Co-operative Bank is formed by federating Primary Agricultural Credit Societies in the district. In each district there is a District Central Co-operative Bank with its branches. They get major part of funds from State Co-operative Bank and provide in turn financial assistance to Primary Agricultural Credit Societies.

The funds for the DCCBs are secured by share capital, loans from the State Cooperative Bank, membership fee from primary societies, deposits from the public etc. The DCCBs were preferred by the general public for depositing their money as they were giving higher rate of interest. However lately due to financial mismanagement in some banks in Gujarat and Maharashtra, general public have started preferring commercial banks to cooperative banks due to doubt about the safety of their deposits.

5.4. Primary Agricultural Credit Society (PACS)

This is at the first (bottom) level of the 3-tier system. Although to start with even 10 members can form a society, presently average membership of each society is 1000 members. The society is located at the village or it serves two or three small villages. The funds of the Primary Agricultural Credit Society is secured through share capital, loan from the District Central Cooperative Bank, deposits of public, membership fee etc. The Society in turn provides loan to its members for various agricultural operations. The loan is given against personal guarantee, mortgage of the property of the borrower and guarantee of the sureties. Being at the lowest level, the entire responsibility of recovery of the loan falls on the society and thus the role of the society is very crucial.

There has been substantial increase in the disbursement of agricultural credit and mostly it is through Primary Agricultural Credit Societies. In 1994, the percentage of loans overdue to loans outstanding was 19.43. The Datey Committee (1974) appointed by the Reserve Bank of India for studying the reasons for high percentage of overdues, has attributed this phenomenon to: faulty lending policies i.e. untimely credit and over financing; lack of supervision; unjustified remission of loans, lack of will on the part of management to enforce recovery or to take strong action against willful defaulters; local unwillingness to develop the culture for timely repayment. There are even now occasions of remission of loans by some state governments and such actions send a wrong signal even to the prompt payers.

The report of the National Commission on Agriculture (1976) states that the benefit of co-operative lending system has mostly gone to bigger farmers and small and marginal farmers could not derive much advantage. The main reasons were stated as: (i) Landownership becoming the dominating criteria both for admission of new members and extending credit; (ii) Co-operative leadership and management were in the hands of bigger farmers who were both unwilling and unable to appreciate the needs of the small and marginal farmers; (iii) The principle of lending being geared to the possible increase in income generation could not be applied to a large coverage of small farmers.

5.5. Land Development Bank

Apart from forming above mentioned cooperative banks for providing short term and medium term credit, Central Land Development Bank at the State level and Primary Land Development Bank at district or taluk level have been established for giving long-term credit. These are also called Land Mortgage Banks. The funds for the Central Land Mortgage Banks are raised by way of debentures guaranteed by the State Governments.

The concept of establishing land development banks originated from the idea of providing loans to farmers for the purpose of debt redemption so as to prevent the land of the poor farmers from being taken away by the *Sahukars* i.e. private moneylenders,

The long-term credit is now generally given for purchasing tractor-trailer or additional land, executing small irrigation work like digging well etc. These banks also continued to provide assistance to farmers for debt redemption. It is important to note that the overdues position of land development banks has been much worse than the cooperative banks and most of the land development banks have become sick.

5.6. Commercial Banks

The main objective of nationalisation of the scheduled banks, which took place in 1969 and 1980, was to give control of these banks to public sector so as to serve greater social purpose i.e. providing services to priority rural sectors etc. In line with this objective, the commercial banks started their operations in the rural area thus supplementing the services of the cooperative banks. The State Bank of India (formed after amalgamation of erstwhile Imperial Bank of India and ten state associated banks) has been in this field from very early stage and the scheduled banks joined the stream subsequent to nationalisation. These banks basically operate on commercial principles and as such provide loan to Primary Agricultural Credit Society or individual farmer after due scrutiny and based on their repayment capability.

Based on the recommendations of Study Group under D.R. Gadgil and the Committee of Bankers under F.K.F. Nariman, the concept of Lead Bank Scheme came into existence in 1969. As per this scheme, districts in the country were apportioned among all the public sector banks and three branches in the private sector. These banks were given the responsibility of preparing the development plans for the areas under their charge. Thus their area of work was enlarged from limiting their activity of providing only credit to preparing plans for meeting the growing development needs. The share of servicing the rural credit by the commercial banks has increased manifold and the co-operative banks are now facing tough competition from the commercial banks.

The commercial banks give short-term, medium-term as well as long-term loans to agriculturists. They give on annual basis direct loans for raising the crops. The agriculturists are supposed to repay the loan at the end of the year so that they can get fresh loan for the next year. They also finance agricultural sector indirectly by way of financing Farmers' Service Societies and Primary Agricultural Credit Societies. The scheme of financing of agricultural sector through Primary Agricultural Credit Societies was started in 1970. This has helped to maintain the functioning of primary credit societies wherever District Co-operative Central Banks were weak.

Through Village Adoption Scheme, the commercial banks were advised to prepare exclusive credit schemes and provide timely and adequate credit to weaker sections of the society.

5.7. Regional Rural Banks (RRB)

In spite of the aforementioned institutional framework it was seen that the benefit could not reach the small farmers and the weaker sections of the rural population. A Working Group on Rural Banks set up in 1975 under M. Narasimhan recommended for establishing new banking institution which combines the major strength of the co-operative credit institution in having the local feel of the rural society and that of the commercial banks in having professional management.

Based on this premise Regional Rural Banks as subsidiaries of the commercial banks have been started since 1975. The composition of the organisation of these banks has been kept such that the staff is familiarised with the needs and the difficulties of the small farmers and the weaker sections of the rural area. These banks are established at locations where cooperative banks and societies were found to be weak. Each bank is meant to cover one to five districts and each of its branches was to finance 5 to 10 Farmers Service Societies.

The Dantwala Committee set up by the Reserve Bank of India in 1977 as well as the Committee to Review Arrangement for Industrial Credit for Agriculture and Rural Development (CRAFICARD) set up by Reserve Bank of India has found the Regional Rural Banks especially suitable for rural developmental works. These banks are being refinanced by NABARD.

5.8. Farmers' Service Societies (FSS)

The institutional set up for delivery of credit served largely the interests of the big farmers and the small and marginal farmers could not reap the benefit. Even where the credit was obtained by the small and marginal farmers, it could not be effectively utilised by them. The National Commission on Agriculture (NCA) in its report stated that, "The central need of the small and marginal farmers is not merely money, but timely availability of a package of inputs and custom services, alone with technical advice and supporting services for storage, transportation, processing and marketing preferably through a single contact point". It is in this context that the Commission, in 1970, recommended the establishment of Farmers Service Societies.

The basic objective of forming Farmers Service Societies as mentioned above is to provide credit along with the extension services to small farmers and weaker sections of the rural population so that the credit is used for the purpose for which it is provided and also it is used productively. There is supposed to be one society for each block. The Working Group on Rural Banks stressed the importance of FSS. The State Governments were asked to attach the FSS to Regional Rural Banks so that the FSS serve as the outlets of RRBs at the base level. The societies started functioning since 1973. However the growth of FSS faced certain problems. Not all states showed interest in forming FSS.

Thus there has been uneven growth of FSS. There happened to be certain amount of overlapping of functions between FSS and Primary Agricultural Credit Societies. Also like other financial institutions, these societies also faced a problem of large overdues.

Apart from the multipurpose credit societies mentioned above, certain cooperatives like sugar co-operatives, milk supply co-operatives have been established to serve the rural people in certain specific areas. These societies provide assistance for not only setting up production facilities but also transportation and marketing. Then as an offshoot of co-operative principle many Self-Help Groups (SHGs) have been established in villages. Women have especially shown lot of interest in running the activities of these groups. The activities include preparing home products as well as taking up small village development projects. Apart from raising the economic level, the formation of these societies has served to reduce the problem of unemployment in the rural area.

5.9. Major Shortcomings of Co-operative Institutions

The co-operative movement and cooperative institutions have played a key role in social and economic development of rural population at a time when India was passing through critical time. The major drawback of this institution continued to be the fact that the weaker sections of the society could not reap the benefit as the rich ones have benefited. Also many co-operative institutions have become sick. Govt. has formed various high-level committees and groups from time to time to study the causes of failures. The common denominator of the findings relate to following areas.

5.9.1. Stratification in Organisation

The management as well as the membership of the societies show the preponderance of rich farmers. Largely these people reap the benefit. Barring a few, others especially weaker sections of society find difficulty in getting the benefit of these institutions. Notwithstanding safeguards provided by the Govt., the changes at the ground level are taking place at snails pace.

5.9.2. Politics in Co-operative Institutions

Certain amount of politics is unavoidable in any institution as it is run by the humans only. However it is found that certain politicians have made co-operatives as their fiefdom and they use them solely to further their political interests. Once they find their way to management positions, they do not allow others to occupy these positions. In many institutions, the same person continues as Chairman or Director of the Managing Committee for 10 to 20 years. Such a situation is not conducive to encouraging the running of organisation on professional lines.

5.9.3. Non-Linkage of Credit with Related Other Activities

The loans are given for a specific purpose. However if the loan is extended without taking into account its related other activity, the purpose of giving the loan may be lost. For example if the loan is given for purchasing seeds but not for fertilizer, the agriculturist may not be benefited. Therefore extension education should accompany the credit to make it effective.

5.9.4. Delay in Sanction and Inadequate Amount of Loan

There are certain cycle of activities in agriculture and all these activities are related to monsoon. Therefore timely availability of loan is very important. Many a times the loan is not given in time due to procedural delays. Also the loan amount is not adequate enough to carry out the whole lot of activity.

5.9.5. Non-Recovery of Loans

Most of the co-operative institutions are becoming sick because effective steps are not taken for recovery of loan. There is also no proper monitoring mechanism for recovery of loans. The waiving of the loan amounts by an executive order of the Govt, save in exceptional circumstances like natural calamities, is counter-productive to a system of recovery of loan.

Some of the shortcomings mentioned above can be overcome by way of adopting a system of Supervised Credit.

6. Supervised Credit

H. Belshaw espoused the importance of supervised Credit. The underlying idea is that small and marginal farmer who needs the credit most also requires certain extension services to use it effectively. According to him, the services have to be extended so as to cover even marketing and allied activities. These services can either be provided by the Govt. or the lending agencies. In a vast country like India, it is difficult for the Govt. to take up this responsibility entirely. They can however take up general awareness and training programmes about extension education. Now a days commercial banks are creating a special cell for agricultural credit and appointing the persons having B. Sc. (Agriculture) to look after the activities of the cell. Thus it becomes easier for them to supervise the use of credit for the purpose for which it was borrowed.

More on this subject matter is given later in the chapter on strategies to be adopted.

Following Figure 3.2 shows number of co-operative societies (credit and non-credit) and membership of co-operative societies from 1990-91 to 1998-99.

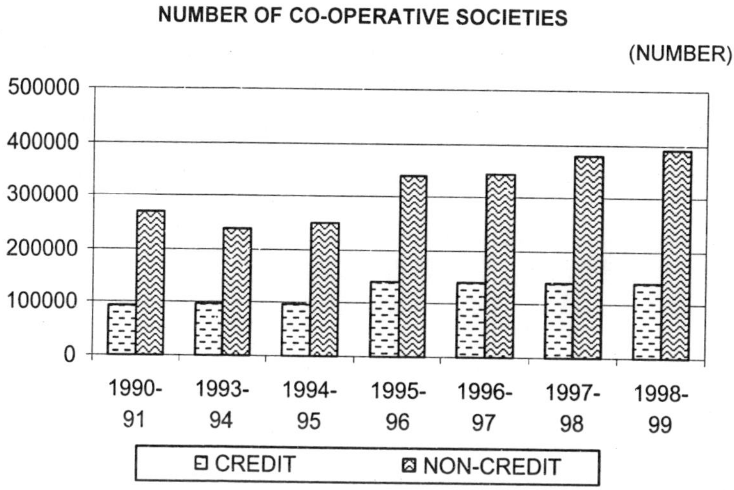

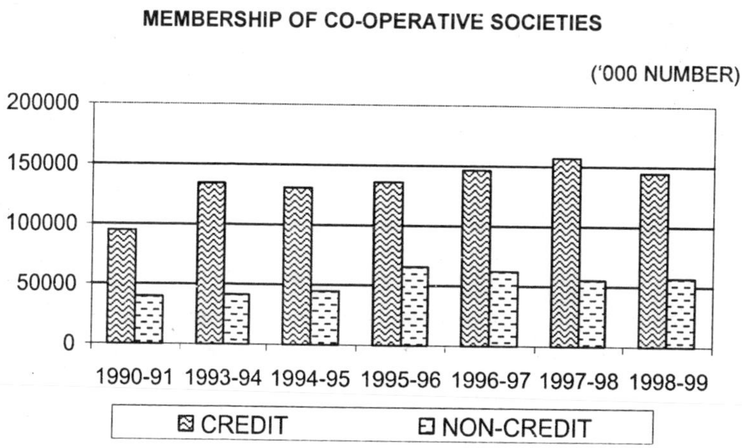

Source: Statistical Abstract India 2003 (GOI Publication).

Fig. 3.2: Number of Co-operative Societies and Their Membership from 1990-91 to 1998-99

9. VILLAGE AND SMALL INDUSTRIES (VSI)

The village and small industries provide maximum employment next to the agricultural sector. Table 3.7 shows employment during 2000-01 to 2001-02 in-village and small industries. Besides employment generation, this sector has certain distinct advantages. It utilises local resources like labour and raw materials for productive purposes. It requires less capital for starting the small industries. As these industries come up in and around villages there is a wide dispersal of industries throughout the region. This is unlike major industries which are located in selected places thus resulting in imbalances in the development of a region as a whole.

Table 3.7: Village and Small Industries-Employment during 2000-01 and 2001-02

(Million persons)

Sl. No.	*Sector*	*2000-01**	*2001-02***
1.	Small scale industries	18.6	19.2
2.	Khadi and Village Industries	6.01	6.27
3.	Coir Industry	0.53	0.54
4.	Handlooms	12.4	12.4
5.	Powerlooms	4.2	4.2
6.	Sericulture	5.4	5.5
7.	Handicrafts	5.7	5.84
8.	Wool Development	1.2	1.2
	Total	54.04	55.15

* Actual

** Provisional

Source: NIRD Rural Development Statistics, 2002-03

Excluding population below the age of 14, and above the age of 60 years, population actually available for work in rural area is 56.3 per cent i.e. 418 million. Thus employment of 55.15 million (2001-02) in village and small industries constitutes 13.2 per cent of the eligible working population in rural area. If not significant, this is not a small figure considering the overall unemployment problem existing in rural area.

1. Khadi and Village Industries (KVI)

Gandhiji expounded that khadi was a symbol of unity of Indian humanity. Khadi may not be soft or elegant in appearance as other mill made cloth nor so cheap but because of simple process and minimum investment required to produce khadi, it has become a source of employment to lakhs of rural people. Further it does suit Indian all weather conditions and it is good for protection of the skin. 6.27 million people are engaged in khadi and village industries. The village industries include handlooms/power looms, handicrafts, coir/ carpet making, milk dairies, poultries/meat, fisheries, (honey) bee keeping, sericulture, food processing, horticulture, soap making etc. Presently food processing, horticulture, dairying, fisheries, meat and poultry are poised for excellent growth. The Govt. proposes to assist village industries by offering certain concessions in taxes and marketing intelligence.

India has certain distinctions in the world such as it has largest livestock, it is largest milk producer and it is second largest producer of fruits and vegetables. India is the second largest producer of fresh water fish after China and fourth largest producer of fish in the world.

2. Small Industries

In order to develop infrastructure for establishment of small industries. Integrated Infrastructure Development Centres (IIDC) are set up. The credit is being provided through Small Industries Development Bank of India (SIDBI). To encourage setting up small industries in the rural sector, the Govt. is providing number of facilities such as land at concessional rates in the earmarked zones, developing infrastructure and offering tax rebates.

Like any industry, small industries are also facing certain difficulties. They have to depend upon raw materials from major industries like supply of yarn from cotton mills to the handloom industry. Globalisation process has also affected this sector unfavourably. In absence of implementation of modern technology and improved productivity, they are finding difficulty in competing with mass production companies. However overheads being low they are still a force to reckon with.

10. NATIONAL SAMPLE SURVEY

10.1. 57th Round (June 2001 to June 2002)

Report on household consumer expenditure and employment situation in India, 2001-02.

National Sample Survey (57th) provides startling facts about the miseries that a large section of population in India is suffering.

10.1.1. Hunger

One household in every 200 in rural India and one out of every 1,000 households in towns go "chronically hungry" i.e. not getting enough to eat in any month of the year. Further sixteen per 1,000 rural households and three per 1,000 urban households are "seasonally hungry" i.e. they do not get enough food in many months of the year. It is ironical that while India is exporting food, its own large section of population lives hungry.

10.1.2. Monthly Consumption Expenditure

The monthly consumption expenditure in rural area is only Rs. 498 as against 933 in cities and towns i.e. rural consumption is almost half of the urban standard. In rural area more than half of the total consumption expenditure is on food.

10.1.3. Shelter

In rural area six per cent of the rural population does not have their own houses. Further whatever houses are there, these are mostly old and dilapidated, built in an unplanned manner and require thorough up-gradation. In urban area six per cent of urban dwellers are either homeless or are living in shanties, 60 per cent live in houses owned by them and 34 per cent have rented dwellings.

10.1.4. Employment

About 60 per cent of the population in villages is unemployed; out of these males constitute 45 per cent and female 69 per cent. In urban area, of the unemployed population, males constitute 45 per cent and females 86 per cent.

10.1.5. Electrification

Only 52 per cent village households have electricity. In urban area too, nine per cent of the households have no electricity.

The above statistics only indicate that whatever has been achieved so far in the field of rural development is not enough. We have to go a long way to bring prosperity in the rural area as well as in urban poor and till then whatever we claim in terms of countrywide good economic performance does not make sense to larger section of the deprived population.

10.2. 58th Round (July-December 2002)-Report on Village Facilities

The National Sample Survey Organisation (NSSO) engaged in collection of socio-economic data employing scientific methods, conducted 58th round from 1st July 2002 till December 2002. The primary objective of this survey was to gather information on social indicators like disability, housing condition, and different types of facilities available in villages. The survey covered 4646 villages throughout the country comprising the states and villages:

> Andhra Pradesh (303 villages), Arunachal Pradesh (71), Assam (226), Bihar (364), Chhattisgarh (84), Gujarat (119), Haryana (79), Himachal Pradesh (100), J & K (95), Jharkhand (130), Karnataka (176), Kerala (158), Madhya Pradesh (240), Maharashtra (291), Manipur (89), Meghalaya (64), Mizoram (35), Nagaland (44), Orissa (190), Punjab (100) Rajasthan (227), Sikkim (56), Tamil Nadu (227), Tripura (102), Uttar Pradesh (578), Uttaranchal (47), West Bengal (370), and Union Territories.

The highlights of the survey are:

- 4646 villages were surveyed in the 58th round of the NSS.
- By 2002, more than three-fourth of India's villages had access to electricity.
- Non-conventional forms of energy were available in less than 12 per cent of Indian villages.
- 55 per cent of India's villages got their drinking water mainly from tube-wells or hand pumps.
- Tap water was used for drinking in about 18 per cent of villages.

- Only 30 per cent of Indian villages had any kind of drainage system; the major types were the open *pucca* and the open *katcha* systems.
- Tube Wells constituted 76 per cent of the irrigation facility in villages.
- Cooperative societies existed in 30 per cent of Indian villages and self-help groups in 24 per cent.
- Government development programmes for provision of drinking water and construction of approach roads were operative in 60 per cent of villages.
- Only one-third of all villages were within 2 km of a telegraph office/PCO/e-mail facility.
- Post offices did not exist in over 78 per cent of India's villages.
- Pre-primary school facilities were available in 66 per cent of Indian villages in 2002 compared to only 39 per cent in 1991.
- 72 per cent of the villages had primary school facilities in 2002 compared to 67 per cent in 1991.
- Literacy campaign/adult education programme was in force in 37 per cent of the villages.
- 54 per cent of the villages were more than 5 km away from the nearest Primary Health Centre and 27 per cent were more than 10 km away from it.
- Only 10 per cent of the villages had a medicine shop and only 20 per cent had a private clinic or doctor.
- Facilities for the disabled such as schools for the blind, deaf and dumb, and the mentally retarded were extremely rare, the situation in this respect being no better than it was 11 years ago.

11. CASE STUDIES

1. Case Study on Development in a Village Natala (Maharashtra)

Situated in an idyllic somewhat hilly terrain, the village Natala is 35 km from the district place and 25 km from the taluk place. The village has 150 houses (10 pucca houses) and a population of 625, with almost equal number of males and females. For drinking water, 2 tube wells have been dug and the hand pumps have been installed on these wells.

Five km long *kachha* road from the main road connecting District place and Taluk place is the only connecting link to the village. While entering the village by this *kachha* road, people have to cross a *nallah.* In the rainy season the water overflows in the *nallah* and the village is cut off from rest of the villages. There is a need to build a bridge over the *nallah* to avoid hardships to the villagers. However the Government has no plans to build the bridge.

The village has 500 acres of agricultural land, 15 farmers owning land more than 10 acres, 25 farmers owning 5 to 10 acres of land and the rest are either small or marginal farmers. There are very few who do not own any land. The land here is fertile for cultivating cotton, soyabeen Jowar, wheat, tur (pulse) etc. There is no perennial irrigation facility. However there are about 25 open wells in various farms catering for partial irrigation. These farmers are growing oranges beside aforesaid produce. The rest of the farmers either could not afford digging wells or they could not find water at a reasonable depth. These villagers have to depend on the rain God. Monsoon is the only source of irrigation for the poor farmers. The villagers also use the water flowing through the *nallah* for irrigation purpose. They feel good about *nallah* at least on this account.

Electricity plays hide and seek game. It is available only for 5 hours a day. Frequent power cuts have become a constant nightmare. During agricultural season, due to no electricity supply during the daytime, villagers have to irrigate the land during night hours without regard to personal safety. There has been an instance of a death of a person due to snakebite. Low voltage is a major problem. Motors of the pumps are getting burnt and the bulb glows feebly. There are 3 TV sets in the village but due to low voltage reception is quite poor. There is a sanction for upgrading the power transformer and stretching new transmission lines but the project could not take off as the cables could not be procured. The village has streetlights.

There is a steady increase in the number of unemployed youth every year. Now these youths numbering 40 out of whom there are 5 graduates make often rounds to a district place in search of a job since agriculture is the only source of employment in the village.

The agricultural work is seasonal, as the main farming season lasts barely for four months. The relief work sometimes undertaken by the Govt. is the only succour to the unemployed youth during off-season. In this context the road construction and building work under Jawahar Rozgar Yojana/Pradhan Mantri Gram Sadak Yojana provide an opportunity for work as well as earning. Unfortunate factor is that the workers are not always paid the prescribed minimum wages by the agencies executing the work.

Till one year back, there was only one telephone connection and it was in the office of Gram Panchayat. Now of course 3 more connections are available. It is fine so long as telephones are in working order. But once there is a fault in the telephone, it takes anything between two to three weeks to rectify the fault, as the repair technician has to come from his place of posting, which is situated at 5 km away.

Most of the small and marginal farmers who constitute the majority are under the debt burden. Those who have taken the loan from the banks are better off. There is however a problem of recovery of loan as banks cannot adopt coercive means as private moneylenders. There are cases when some villagers took loan from the private lenders, they had to pay heavy interest and in some cases even crops were mortgaged to the lenders.

The most vulnerable section is small and marginal farmers. This section does not afford digging wells, purchasing water pumps and payment of electric bills. Agriculture is still carried on in a conventional way and modern technology is still a far cry. One bad monsoon makes things difficult and two consecutive bad monsoons make things worse. It all happens due to abnormal increase in cost of inputs such as fertilizers, seeds, insecticides etc and on the other hand there is no proportionate increase in the prices of produce. On the top of this if there is a bad monsoon, farmers run into losses. Hence farming has lost charm, which once upon a time it carried. 60 families are in below poverty line category. Some farmers have diversified their activities in cattle rearing and business of milk. Such farmers are relatively better off.

Many of the villagers are unaware of the Government's welfare schemes like social security, old-age pension, kisan credit card, financial aid for girl's marriage, scholarship for education, special schemes for below poverty line families etc. Under National Old Age Pension Scheme, pension is Rs 250 pm. Fortunately two persons in the village are getting the pension. Some of the villagers did avail facilities of some of the Govt. schemes like construction of toilets, drains, low-cost houses for the poor. There are two Self Help Groups, one organised by the women and the other by the young boys. They organize savings from among the members, prepare home based products and undertake small works which fetch them some money.

There is a separate Gram Panchayat office building. The institution of Gram Panchayat has given the villagers a sense of empowerment especially women and scheduled castes and scheduled tribes who constitute minimum 50 per cent of the total membership. However there are occasions when husbands instruct their wives to take particular decision even when wives are the members and husbands are not. On revenue side, the Panchayat collects property and water taxes. But the amount collected is so small that the Panchayat has to depend mostly on Govt. funds. Although the village has a small room meant for the library but there are no books. Public Distribution System outlet lacks the availability of enough stock of foodgrains.

None of the teachers except one posted in the village stay back. They commute from the nearby town and leave the village as early as possible.

There are two *gobar* gas plants but these are not in working order. There are 10 motorbikes and 25 ordinary bikes.

Unfortunately some of the adults indulge in drinking liquor which is not liked by their counterpart womenfolk, but they are helpless before the men. The country liquor is available locally. Those who indulge in drinking liquor believe that this is the best way to forget their worries borne out of abject poverty.

To sum up, most of the families in general still lack the kind of minimum standard of living. The landless people work as labourers so long as they have enough strength and the work. In old age after their savings get exhausted, they feel they have nobody to look after and they are at the mercy of their children,

Hard life and drudgery in the village with no hope of getting better is compelling many in the village-especially youth to seriously try for migrating to towns/cities. Already some have succeeded in their attempt.

2. Case Studies of Input Costs and Output Prices of Agricultural Produce

Two case studies have been conducted by the author at village Natala in the year 2002-03 regarding input costs and the income to arrive at the net income to the agriculturist. One case study pertains to a cotton and the other pertains to Soyabeen.

CASE STUDY-I

Case Study of Cotton Production in village-Natala (Maharashtra)

Land under cultivation:	**7 acres**
Sowing period	3rd week of June 2002
Harvesting time	November 2002 to January 2003
Cost of production:	**Rs.**
Farm cleaning	1000/-
Tillage	4000/-
Seed	7700/-
Fertilizers	5600/-
Sowing	3000/-
Interculture	5000/-
Insecticides and Spraying	4000/-
Picking/Harvesting	6000/-
Irrigation power charges	3000/-
Rozdar (Helper)	6000/-
Sub-Total	45300/-*
Interest on the Investment @ 8.5 per cent for 9/2 months	**1500**
Total	46,800
Yield	27.5 Quintals
Income	Rs. 2200/- per quintal; Total Rs. 60,500/-
Total Expenditure	Rs. 46,800/-
Net income	Rs. 13,700/-

* The expenditure of Rs. 45300/- does not take into account the supervision charges of the farmer.

CASE STUDY-II

Case Study of Soyabean Production in Village-Natala (Maharashtra)

Land under cultivation:	**10 acres**
Sowing period	3rd week of June 2002
Harvesting time	November 2002

Cost of production:	**Rs.**
Farm cleaning	1500/-
Tillage	5000/-
Seed	9400/-
Fertilizers	4950/-
Sowing	4000/-
Interculture	3000/-
Insecticides and Spraying	3000/-
Cutting of crop/Harvesting	5000/-
Irrigation power charges	2000/-
Rozdar (Helper)	4000/-
Sub-Total	41,850/-*
Interest on the Investment @ 8.5 per cent for 6/2 months	900/-
Total	42,750/-
Yield	44 Quintals
Income	Rs. 1350/- per quintal; Total Rs. 59,400/-
Total Expenditure	Rs. 42,750/-
Net income	Rs. 16,650/-

* The expenditure of Rs. 41850/- does not take into account the supervision charges of the farmer.

The above case studies pertain to the land where irrigation facility from open well is available and the monsoon has been more or less normal. In spite of this it is seen that the ultimately the net income from one acre is Rs.1957/- in the case of cotton and Rs. 1665/- in the case of soyabean. In the case of soyabean, it has been possible to take the second crop of wheat. This fetches another Rs. 700/- per acre. In the case of soyabean plus wheat, the income is Rs. 2365/- per acre. Thus if a farmer has 5 acres of land, the average income for the year would be 10000/- i.e. less than Rs. 1000/- per month.

In the above calculations, the supervision charges of the farmers has not been taken into account. The above case studies represent the general trend that exists in the rural area. In fact the condition of marginal farmers is still bad. Thus there is a distressing situation when the monsoon is unfavourable in any year.

The above case studies lead us to a reason as to why even after 57 years of Independence the lot of agriculturists has not been improved. It also indicates the necessity of taking problems of rural society more seriously. If we look at only from the bare input costs and output income, it becomes necessary not to allow any further increase in the cost of fertilizers, insecticides and power. At the output end, the farmer must get the remunerative price. Therefore when there is a debate going on in the country about removal or reducing the subsidy on inputs to the agriculture, the impact of such an action should take into account the present economics of agriculture. While reasonable subsidy on supply of power to agriculture is fully justified, the suggestion of giving free power to agriculture will not be in the long-term interest of the farmer as discussed in a chapter on strategy.

12. STRATEGIES TO BE ADOPTED FOR AGRICULTURE AND RURAL DEVELOPMENT

While the achievements due to the efforts of the Govt. and Non-Govt organisations in rural development are not small, the fact remains that the rural area still accounts for more than nearly 75 per cent of the country's below poverty line population. In terms of basic education, primary healthcare, housing, safe drinking water, sanitation and proper infrastructure, there is a long way to go in providing these minimum needs in rural area. It would be impossible to achieve the goal of providing these minimum needs in normal course and there is a necessity of adopting certain strategies. These have been briefly touched upon earlier at the appropriate places where various Programmes/Schemes have been discussed. However main strategies are discussed below:

1 Basic Facilities

1.1. Housing

There are quite a number of landless labour in rural area who are still without housing facilities. Further, in olden times, houses were built in half hazard manner and these require up gradation. Indira Awas Yojana (IAY) has been the main instrument to provide housing to Scheduled Castes and Scheduled Tribes as well as to the non-SC/ST rural poor. In order to complement IAY, the Golden Jubilee Rural Housing Scheme was launched in 1997 to give a boost to rural housing. A major impediment to credit for rural housing is identified as absence of proper title to the land. The Govt. of West Bengal has made a law to simplify the creation of security. This law with modifications if necessary deserves to be emulated by other states.

Funding for the housing programme is also provided through Housing and Urban Development Corporation (HUDCO) and General Insurance Corporation. Since these institutions are saddled with many other sectors, there is a need to strengthen the present institutional set up and monitor closely the utilisation of funds. There is a need to set the target of 2.5 million housing units per year to clear the hugebacklog. For this to happen, there is a need to reduce the rate of interest on the loan and enhance the allocation of funds as the allocations made so far have been found to be quite inadequate to meet the targets fixed.

1.2. Education

The results of the past as well as present literacy drives in the elementary education in the age group of 6-14 years are far short of the targets set from time to time. The literacy percentage in the girls is especially very low. Similarly there is a need to expand adult education. In view of the cost involved there is a shift towards introducing non-formal education. Instead of recruiting new teachers in regular category, *Shiksha Karmis/para-teachers* are being recruited on contract basis. Due to this quality of education is obviously affected. However looking to the enormity of the population to be covered there is no alternative but to supplement the formal education with the non-formal system. Main problem is found to be lack of punctuality and attendance of the regular teachers most of whom are living at outstations. These have to be ensured by regular inspections from the administering departments. However since this system could not succeed, it is better that this task is given to Village Committees. As regards para-teachers, there is a need to give at least short-term orientation training before they are inducted for the teaching.

All aspects pertaining to education in rural area have been dealt in detail in Chapter-2 of this book.

1.3. Healthcare

Integrated healthcare comprise healthcare education, medical treatment, provision of clean drinking water, sanitation etc. Three tier system of healthcare i.e. establishment of district hospital (DH), primary health centre (PHC) and sub-centre (SC) has been put in place but there are still number of villages which are away by more than 5 to 20 km even from the SCs. Besides, PHCs and SCs are poorly staffed and even essential medicines are not available in time. These centres are also poorly equipped. Doctors are neither self motivated nor conditions in the centres are encouraging to them professionally. While there is no doubt that financial crunch is responsible to a great extent for such a pathetic state of affairs, bureaucratic approach of the health authorities and the health ministry are equally responsible for not addressing the problems which the PHCs/SCs are facing such as timely posting of staff, proper accomodation to them, timely availability of medicines and funds etc. It is therefore necessary to post medical/para-medical staff who are willing to work in rural area and staff at the administrative offices should be more sensitive towards providing timely inputs to PHCs/ SCs. Village Health Guides are supposed to create health awareness and provide health education and thus augment health services in the rural area. Here again VHGs should be given orientation training before induction so as to make the system successful.

All aspects pertaining to healthcare in rural area have been dealt in detail in Chapter-4 of this book.

1.4. Nutrition

Nutrition aspect should be taken care of by way of providing looked mid-day meals for the school children instead of providing raw food articles. There is a need to bring awareness about the importance of nutrition in the female population especially the nursing mothers.

A National Food Guarantee Scheme which is on the anvil would focus on availability of food as well as quality of food that would be available in the country especially in rural area. Planning Commission has chosen 150 villages as part of the food programme in rural area, and a seven-point action plan has been outlined to achieve the "hungerfree" goal by 2007. Under the programme, over 50,000 farm schools would be set up all over the country that would encourage farmer-to-farmer learning. There are plans of setting up 'community food banks' by local self-help groups in chronic under and malnutrition regions. Special focus will be on delivery of nutrition support programmes for pregnant women and extending up to old and infirm people, giving particular focus to HIV/AIDS and tuberculosis patients, infants below 2 years and adolescent girls and nursing women. A nutrition literacy programme include providing every household with an "entitlements card" indicating their entitlements to the nutrition safetynet. However it is to be seen how effectively the above programme works at the ground level.

1.5. Clean Drinking Water

Although the State has taken up the programme of construction of tube wells for providing clean drinking' water in the villages, there are still many villages where it is not available in their own villages or even within a reasonable distance. The State has to accelerate its efforts in providing this basic necessity. In absence of clean drinking water, the villages often face epidemics like cholera, gastroenteritis etc.

There are many tubewells after working for sometime become out of order. There is a need to repair these by engaging suitable agencies.

The norms being adopted for providing drinking water to rural population in the habitations are 40 litres per capita per day (LPCD), 30 LPCD additional for cattle in Desert Development Programme (DDP) areas, one hand pump for every 250 persons, potable water sources within 1.6 km in plains or 100 metres elevation in hilly areas. In view of the gigantic task involved to provide drinking water as per norms, GOI has now ordered a fresh nationwide survey to enable it to take a comprehensive view. Besides drinking water supply, the emphasis has to be on quality of water as well.

1.6. Sanitation

Rural people are well recognised for maintaining highest cleanliness in their own homes but the same standard is not followed when it means throwing personal garbage or drain outlet on the public lanes or roads. Since olden times there was no system of simple and cheap toilets in vogue and villagers developed habit of using road side area for attending to nature's call. Only 6 per cent of households in the rural area have household toilet facility and the others go for open defecation mostly alongside the road or street. Women therefore prefer the timings either before sunset or after the sunset. Now States have taken up construction *of Sulabh Shouchalaya* costing Rs 2000/- each for which beneficiary has to pay Rs 500 only and the balance amount is borne by the State. This scheme has met with partial response, probably due to the reason that rural people find it difficult to pay even Rs. 500/- for the toilet. Also some of the villagers are wrongly using newly constructed toilets for storing materials, obviously due to no other place being available for storing materials. It is necessary to increase the subsidy and cover every household with *Sulabh Shouchalaya* in rural area and at the same time educate villagers about the importance of sanitation so that it is used for the intended purpose. The village panchayat or the village committees can best ensure this.

Although healthcare should comprise both preventive aspect as well as treatment of disease, emphasis has to be laid on preventive care in the rural area. Malaria, gastroenteritis and smallpox are some of the common diseases in rural area. In this context providing safe drinking water and sanitation facilities assume great importance. The Primary Health Centres (PHCs) can play proactive role in providing medical care as well as explaining awareness about maintaining proper sanitation.

2. Infrastructure Development

2.1. Rural Connectivity

2.1.1. Roads

Among the most important infrastructure needs of the rural area, roads occupy the important position. Presently, the conditions of the roads connecting cities-district places-taluq places are only satisfactory, as maximum attention is paid to them. As regards rural area, many of the villages are still not accessible by roads. Even wherever roads are provided these are not all-weather roads and these are generally poorly maintained. State can take up development of roads work in the rural area especially during lean agriculture season. This will serve twin purposes i.e. development of rural infrastructure and also giving employment to rural population during much needed period.

Looking to the fast deterioration of the roads in the rural area, there is a need to have a proper quality check to ensure that the work is executed as per the specifications, and if it has not been executed as per specification, action should be taken against the erring staff.

2.1.2 Communication

The telephone communication is yet to reach many villages in India. The Bharat Sanchar Nigam Ltd. has over 30,000 exchanges conferring to a band width connectivity to nearly 80 per cent of the country. If these cater to 20 villages each, all India's villages can be covered soon. However only laying cables and providing telephone sets would not suffice. The installation of telephones has to be accompanied by provision of satisfactory maintenance facilities for attending to faults, in reasonable time, in the system.

2.2. Minor Irrigation/Watershed Management

Due to land ceiling act, the size of land holdings has got significantly reduced. The holding per member got further reduced on account of large size of families (in absence of adequate response to family planning) and further divisions in the combined families from generation to generation. This has resulted in 80 per cent of the farmers falling in the category of small and marginal farmers. Such holdings are not economically viable unless some kind of irrigation facility is provided. Major irrigation projects involve a lot of financial investment, displacement of large population, submergence of large tract of land and long gestation period. As such it is becoming increasingly difficult to take up major irrigation projects.

Through the ages, Indian agriculture has been sustained by natural and man-made water structures such as lakes, tanks, ponds etc. The emphasis has now to shift from major irrigation projects to minor irrigation and better watershed management. These can be in the form of open well, nearby water tank or minor irrigation canal. There are success stories in watershed management from places like Ralegaon Siddi and Hiwre Bazar in Maharashtra and organisations including Tarun Bharat Sangha in Rajasthan. As in the case of roads, these works can also be taken up in lean agricultural season so that twin purposes i.e. development of irrigation facilities together with employment to unemployed as well as idle agricultural labour are achieved. The State can play important role by advancing cheap credit, planning such works in advance with due publicity and involving NGOs and local communities. On completion, the Panchayat Raj Institutions can be encouraged to own, operate and maintain these water structures.

2.3. Electricity

Access to Electricity: Due to preponderance of marginal and small agricultural land holdings, some kind of irrigation facility has become a necessity to make these holdings viable. Such irrigation facility needs pumps which operate on electricity. Electricity is also required in villages for lighting purpose. State Electricity Boards (SEBs) have been given the responsibility of supplying electricity to the villages. To give a thrust to electrification in rural area Rural Electrification Corporation has been set up, which assists SEBs in funding the programme. According to the new definition of an "electrified village", a village will be considered electrified, only if a minimum of 10 per cent of households in that village have power supply. There are still 1,25,000 villages (80,000 as per earlier data) to be electrified. It is planned to connect them through Accelerated Rural Electrification Programme in phases. This will take at least 5 years.

Table 3.8 shows the number of towns and villages electrified as on 31.03.2003 Since last few years there has been slow down in rural electrification in view of the poor financial health of the SEBs. Further there is incidence of greater degree of load shedding in rural area because electricity fetches lesser revenue from agriculture. It is often cited that like industries, electricity should be used in agriculture during off peak hours i.e. during nights, oblivious of the fact that it is quite unsafe to work in fields during nights. There are occasional reports of some people having got bitten by the snakes.

In view of the above, there is a need to accelerate the programme of rural electrification. According to the National Electricity Policy - 2005 approved by the GOI on 2nd February 2005, every household will be provided with electricity by 2010.

Table 3.8: Number of Villages Electrified in India as on 31.03.2003

State/	*Villages*		
	Total (1)	*Electrified (2)*	*Percentage*
State:			
Andhra Pradesh	26586	26565	100.0
Arunchal Pradesh	3649	2316	63.5
Assam	24685	19039	77.1
Bihar	67513	48048	71.2
Chattisgarh	19720	18821	92.9
Goa	360	360	100.0
Gujarat	18028	17940	100.0
Haryana	6759	6759	100.0
Himachal Pradesh	16997	16890	99.4
Jammu & Kashmir	6477	6300	97.3
Jharkhand @	—	—	—
Karnataka	27066	26767	98.9
Kerala	1384	1384	100.0
Madhya Pradesh	51806	50400	97.3
Maharashtra	40412	40351	100.0
Manipur	2182	2007	92.0
Meghalaya	5484	2757	50.3
Mizoram	698	691	99.0
Nagaland	1216	1216	100.0
Orissa	46989	37307	79.4
Punjab	12428	12428	100.0
Rajasthan	37889	36885	97.4
Sikkim	447	405	100.0
Tamil Nadu	15822	15822	100.0
Tripura	855	817	95.6
Uttranchal	15681	12863	82.0
Uttar Pradesh	97122	55230	56.9
West Bengal	37910	31367	82.7
*ALL INDIA	587258	492325	83.8

* Included Union Terrotories (1) As per Census 1991 (2) As on 31.03.2003 @ Included in Bihar

Source: Statistical Abstract, India, 2003 (GOI Publication)

2.4. Non-Conventional Energy Sources

The renewable energy started getting importance in India from early 1970s. India has today one of the largest programmes for renewable energy. The activities coyer all renewable energy sources, such as biogas, solar energy etc. These sources have special significance for remote habitations as there are practical problems in providing electricity and other energy sources to such habitations.

2.4.1. Bio-Energy Sources-Biogas

So long, wood was being used as a fuel for cooking food. However due to indiscriminate cutting of trees in forest area and social forestry being inadequate to meet the requirement, wood is not generally available for cooking.

Bio-gas produced from animal dung has been found to be very simple and cheap source of energy for cooking, heating water etc. This process of using dung for energy has also improved sanitation in the rural area. The State should continue encouraging setting up bio gas plants on a large scale by giving adequate subsidy as it is beyond the capacity of rural poor to do so. Biogas can also be used for lighting purpose where electricity is not available.

There are number of bio-gas plants which have been installed with initial fanfare but are not in working order. The common complaint is that once the system goes out of order, the agencies that have erected these plants are nowhere to be seen to rectify the defects. Mostly the defect noticed is that due to improper alignment or inadvertently falling of some stony material inside the well, the large metallic lid gets stuck on the periphery and does not move. There should be a crash programme to remove such technical defects and put all biogas plants in working order. Another reason for lying these plants idle is that the users themselves abandon their use either due to lack of interest or non-availability of enough cow dung. In this case it is useful if some nearby families join together for installing and maintaining the plants.

2.4.2. Bio-Animal Energy

Bio-Animal Energy is another corollary to Bio-Energy. Notwithstanding increasing use of mechanised equipments like tractors etc in agriculture, farmers still depend largely on age old system of animal energy of which bullocks constitute major portion. In order to get better output, there is a need to improve breed of the animals. Since agricultural work is seasonal, animals are fed well during working season and for the remaining period the animals have to be contended with grass in the pastures if available. The animals are thus underfed and in some cases one gets a feeling that this amounts to cruelty to them. It is therefore necessary that for better output the animals are fed properly round the year. In this context, a new "Livestock Food Corporation" to be set up is expected to devote itself to fodder and feed banks.

2.4.3. Solar Energy

Solar Energy is being utilised through two different routes, namely, solar thermal (ST) and solar photovoltaic (SPV). The technology for the manufacture of the SPV cells and modules has been developed almost entirely on the basis of domestic R&D. The module production in India to the tune of 23 MW accounts for 5 per cent of global production.

The solar thermal devices are being utilised for water heating, space heating, cooking, drying, water desalination, industrial process heat, steam generation for industrial and power generation applications etc. The solar photovoltaic system is useful for applications such as lighting, water pumping and telecommunications and for producing electricity to meet the needs of the villages.

Notwithstanding above advantages, the response to use of solar energy devices is lukewarm. This is mainly because the initial cost of installation is more in comparison to conventional sources in spite of the fact that Govt. offers considerable incentives. Presently conventional electric power has not reached in remote habitations. It will be easier to cover these habitations through SPV system. Once properly installed, community can be involved in managing the system for which necessary training should be arranged. Govt. can play more proactive role in this respect.

2.5. Storage Sheds/Warehouses and Cold Chains

Foodgrain production in India has touched 212.9 million tonnes but storage capacity is only 70 million tonnes. When new produce arrives in the market, the prices are low. They start rising after about two to three months. Since farmers generally have only small houses for living purpose, they need facility of safe intermediate storage. These storage sheds have to be nearer to the producers, as transportation to a longer distance will involve higher costs. The difficulty of transportation is further compounded by the poor conditions of the roads in the rural area. The perishable items like vegetables, fruits' and milk require cold chains to avoid their spoilage. The availability of proper packaging facility is also important in case the produce is to be sent to long distances. These facilities acquire all the more importance if the produce is to be exported.

3. *Human Resource Development (HRD) & Organisation Structure*

3.1. Human Resource Status

Unlike usual thinking that large population is a drag on the development of the nation, it is opined that India's vast population can turn out to be a great strength provided their potentialities are tapped properly. Empowering the people with universal access to education, health and facilitating their full participation in the growth process through gainful employment will not only enhance their welfare but will also reinforce the overall economic growth process. In this context, 72.22 per cent of the total population and 58.4 per cent of the country's total labour force that is living in the rural area can also turn out to be the big resource for the country if they are trained properly. Of this, excluding 35.7 per cent of the population within the age group of 0-14 years and 8 per cent of the population of the age of 60 years and above, rest of the population (56.3%) is actually available for work. Further, active employment in agriculture is seasonal. It is therefore necessary to provide (i) Extension education to farmers so as to enhance productivity of the agriculture. Krishi Vigyan Kendras can play important role in this and (ii) Training facilities to rural artisans which will enhance their potentialities to be engaged in the rural industries.

Vocational training in Industrial Training Institutes (ITIs) can perform the function of such training. The skills imparted by ITIs must keep pace with the technological demands of industry. Govt. now proposes to launch a programme in Central Sector to upgrade ITIs by way of providing appropriate infrastructure, equipment and upgrading the syllabi.

As mentioned earlier, since the work in agriculture is seasonal, some of the states have started schemes to generate self employment. The schemes include setting up small-scale agro-based industrial units. The State provides 25 per cent of the project cost and the rest is to be borne by the beneficiary. However due to paucity of availability of even funds to the extent of 25 per cent with the farmers, these schemes have generated little response. It is therefore necessary to enhance the State's share to say 40 per cent.

3.2. Employment

In rural area, the main avenue of employment is agriculture. Village and small industry sector is another avenue. The typicality of the employment in agriculture is that its activities are season oriented and the period during which major agricultural activities are carried out is generally 4 months in a year. Therefore prevalence of underemployment in rural area is predominant. This is the main reason for a shift of surplus labour from rural to urban area, rural youth going to towns/cities in search of jobs. The solution is taking up, during lean season, developmental works such as constriuction of roads, building ponds/water tanks, taking up minor irrigation projects etc and engaging local rural unemployed persons on these works. The schemes which provide work oriented towards building permanent assets as well as remuneration to workers should be encouraged.

Some states have started schemes of giving unemployment allowance to those who are registered in employment exchanges but could not get the employment.

For example, Maharashtra State started Employment Guarantee Scheme (EGS). Unfortunately many a times those working under EGS are either paid less than minimum wages or do not get their wages in time. The National Rural Employment Guarantee Bill 2004 has been introduced in Lok Sabha on 23rd December, 2004. The Bill seeks to provide for the enhancement of livelihood security of the poor households in rural areas of the country by providing at least one hundred days of guaranteed wage employment in every financial year to every household whose adult members volunteer to do unskilled manual work. The bill is stated to be on the lines of Employment Guarantee Scheme (EGS) in Maharashtra and that the experience gained in Maharashtra and the pitfalls encountered in EGS have been taken into account while drafting the bill.

Needless to say, to give unemployment allowance is all right as a short time measure in circumstances like drought, any natural calamity or some specific contingency. However efficacy of such schemes on long term basis is yet to be proved and there is every possibility of these schemes actually turning out to be counterproductive in the sense that the main focus on finding permanent solution may get shifted to taking recourse to half hearted measure which is tantamount to under employment. If continued for a longer time, persons may loose their initiative in making sincere efforts in self-development and finding a proper job. The permanent solution lies in making higher investment in agriculture and rural infrastructure, providing incentives to village and small industries, encouraging self-income generating schemes and stimulating overall growth of rural economy resulting in creation of assured job opportunities.

3.3. Education and Vocational Courses/Training

Literacy rate in rural area is barely 60 per cent. Education on account of its importance has been made a fundamental right in the Constitution. To make every child literate between the age group 6 to 14, the State has to mobilise enormous resources for providing school staff, buildings, teaching aids, drinking water, toilets, play grounds etc. As the system is presently directed towards enrolment of more number of children, the quality of education has been relegated to second priority. Obviously there is high percentage of dropouts at each level i.e. Fifth, Eighth and Tenth level. What will these youths, with no practical experience, do when they leave the schools? Gandhiji visualised such a situation and therefore emphasised to include vocational courses in the curriculum. After school education, the Industrial Training Institutes and Polytechnics can offer specialized vocational training and Rural Institutes can work for capacity building to undertake extension programmes in agriculture & allied services and village industries. These have to remain thrust areas in coming years for empowering rural youth.

3.4. Empowerment of Panchavat Raj Institutions (PRI)

All India Conference of Heads of Panchayats from across the country was held in May, 2003 at Delhi to review the functioning of the PR system. It was highlighted in the conference that the functioning of Panchayat Raj Institutions (PRI) has been affected in the absence of adequate fund made available to them and its proper utilisation. In the National Declaration which was adopted at the conference, it is proposed that the money sanctioned for Panchayats will hereafter go to them directly, but via the State Govt. with the instruction that the money earmarked for Panchayats has to be used exclusively for the development purpose. It is stated in the proceedings that the proposal is borne out of a concern over "ineffective functioning" of Panchayat Raj Institutions in the absence of adequate fund and its proper utilisation. However national consensus will have to be built for this proposal to be implemented as State Governments should be willing to accept the decentralisation, since ultimately monitoring of utilisation of funds by PRIs will have to be done by the State Governments only.

3.5. Role of Village Leadership

Village leadership plays a pivotal role in the process of development of the village. This can be observed from the variations in the extent of development in similarly situated villages but with different leadership. The degree of development of the village having committed leadership is far better than the village having indifferent leadership. There are number of schemes which have been established by the State for the benefit of the villagers. However there are lacunae in the implementation of these schemes, important one being there is no full awareness in the community about the schemes. Village leadership can play vital role in this respect and also assist those villagers, especially uneducated and poor, who are not conversant how to reap the benefit from these schemes.

In our democratic set up people have allegiance to different political parties. This system in the context of a village sometimes works to the disadvantage of the villagers in the sense that the system results in creating artificial barriers in the small village population. Since villagers' dependence on nature is already high and there already exists division on caste lines, they can ill afford in creating another manmade hurdle in development of the village. In this context, the role of the village leadership becomes all the more important. The leadership must rise above petty party politics and stand united on the issues concerning development in their village. Political parties also should display their maturity in allowing village institutions to function without any outside interference.

3.6. Promoting Family Welfare Schemes

As per 2001 Census, India's population is 1028.7 million and census figures suggest that by 2035, India will have a staggering population of 1.46 billion, far outstripping China? Clearly our family planning and reproductive and child health policies have not succeeded as expected. The Family Welfare Programme throughout the country has resulted in reduction of birth rate more in urban area than in rural area. Large family size has always been a bane of abject poverty in the rural area. Even now early marriages in rural area is the norm. Especially in tribal families, women in their early 20s are mother to two children. The campaign of awareness of the importance of small family norm has somehow not yet met with desired success in the rural area. The innovative ways such as dramas, mobile film shows, programmes on TV/Radio, distribution of literature etc. on this subject can bring more awareness. Gram Panchayats/ Gram Sabhas and NGOs can play important role in this area and they should be fully associated in the family planning campaign. Promotion of education especially in girls is however the best way to make Family Welfare Programme successful.

3.7. Role of Women in Rural Development

As per 2001 census, rural population is 743 million as against India's total population of 1028.7 million and the female population is 48.61 per cent of the total rural population. The role of women is therefore very crucial for rural development. Indian civilization has a chequered history. It is apparent from the scriptures of the Vedic period that women were well respected in the society and there was no discrimination on the basis of gender. The discrimination somehow started with male domination in the society and perpetuated later on. However Pre-Independence era saw many reformers like Raja Ram Manohar Roy, Gopal Krishna Gokhale and Gandhiji who worked hard for emancipation of women and making them equal partners in the development process. Provisions have been made in the constitution with regard to safeguarding the interests of the women. In spite of this, they are still far off from becoming equal partners of socio-economic-political system. Irrespective of the sharing of work pattern and family income, the distribution of food, education, health, shelter space or decision making in the family is significantly skewed in favour of men and against women. Education of girls thus remains to be the major thrust area to improve the situation. The literacy rate in girls in the rural area is barely 47 per cent. The rest of them are engaged in the family chores at home or in the agriculture and related areas. The progress of their enrolment in schools is quite slow because of less priority given by the parents for education to girl child and prevalence of utter poverty in the rural area.

As regard women's participation in work, the table 3.9 shows occupationwise female/male workers in rural area in 1991 and 2001.

Table 3.9: Occupationwise Female/Male Workers in Rural Area in 1991and 2001

(in percentage)

Sl. No.	*Category of Workers*	*1991*			2001		
		Persons	*Male*	*Female*	*Persons*	*Male*	*Female*
1.	Cultivators	48.2	51.4	38.6	40.1	42.2	36.5
2.	Agricultural labourers	32.2	26.4	49.3	33.2	27.5	43.4
3.	Others workers	17.5	20.3	9.2	26.7	30.3	20.1

Source: NIRD Rural Development Statistics, 2002-03 and Census of India.

It is seen from the above figures that the participation of women in the cultivator and agriculture category is equal to the men. Also with time there is a clear shift of surplus labour from cultivator and agriculture category to other occupations in the rural area.

It is pertinent to note that there is a difference, though unjustified, in daily wages of men and women even for the same work in the agriculture, i.e. female gets less wages than her male counterpart.

With a view of raising potential of women, organisations by name Mahila/Bhagini Mandals have come up in several states and they are helping women a lot. These Mandals are working towards enhancement of skill and exploring different avenues of utilising women's creativity. The Indian Council of Social Science Research (ICSSR), while appreciating the work being done by the Mandals have however noted that the Mandals lack adequate representation of weaker sections of society and scheduled casts in the organisations. Women have also been recognised as a target group for development under IRDP/TRYSEM. Further, organisation named Development of Women and Children in Rural Areas (DWCRA) was set up in 1982-83 for providing training for women in rural crafts, weaving, painting, dairy development, poultry, honey bee keeping etc.

While Self Help Groups (SHG) through empowerment of women have come up in various states, in Andhra Pradesh there is a massive participation of women in these groups. SHG has 16 to 20 members and assuming average family size of 5-6 members, each group serves 100 persons. Development of Women and Children in Rural Areas (DWCRA) provides loans to these groups. In turn, the group gives loans to individual members and ensures its repayment by persuasion and if it fails, by using social pressure. Looking to the excellent repayment record, NABARD, the rural lending wing of the Reserve Bank of India started in 1992 financing SHGs through formal banking system and used NGOs as facilitators. They introduced a rating system for SHGs to emphasise the importance of creditworthiness. Witnessing the performance of these groups. World Bank also came forward to finance a programme which now enabled SHGs to hire skilled persons and create Community Investment Fund for taking up small infrastructure work. Matured SHGs may eventually be in a position to graduate from consumption or production credit to starting micro-enterprises. The institution of SHG is not merely to get the finance and use it for needs of the members but it represents much more than this. It represents empowerment of women so as to enable them to participate in developmental activities.

As per the Hindu Succession Act, the woman is entitled to an equal share as her male siblings in her father's property but has no right to ancestral property, Since family law is one the concurrent list, as a measure of social reform, amendment to the Hindu Succession Act has been introduced in Andhra Pradesh (in 1985), Tamil Nadu and followed by Maharashtra (in 1994). The amendment seeks to give equal status to daughter in the matter of ancestral property of HUF. Instead of each state drafting its own separate legislation granting equal status to women, the central government should come out with a bill applicable for the entire nation however, there is a word of caution here.

No society can claim itself to be a liberal democratic when women are denied their just property rights. The amendments are bound to create a sea change in women's own perception of their worth. But the amendments are just the first steps towards empowerment of women. For a positive result, there has to be a positive change in the social mores.

While women's role in social and cultural activities is significant, their presence in developmental projects is still insignificant. It is therefore necessary to redouble the efforts being made for enhancing the status of women in the development process.

3.8. Participation of Non-Govt Organizations (NGO) and Corporate Sector

The field of Rural Development is so vast that the efforts of Govt. alone are not going to be adequate. To achieve the goal, it calls for multidimensional and multilevel approach. In this context NGOs can play important role. They can bridge the gap between the State Organisation/Credit Institutions and the rural people. They can get the funds from any source i.e. private or public. Since they are not directly controlled by the State, they can adopt innovative ways to solve the problems being faced by the rural people. However Govt. should monitor performance of those NGOs who are receiving funds from the State, as not all NGOs are sincere and committed in as much as not all are bad. If performance of any NGO is found to be not satisfactory, the State should derecognise such an organisation for the purpose of funding. There are also reports that some the officials who are dealing with NGOs extract money from NGOs with a promise that the officials will give favourable report so that the NGOs can continue to get the Govt. aid. In such cases the performance of the NGOs is affected. Therefore there is a need to adopt transparency in the working of the Govt. system dealing with NGOs. Adopting a Committee system instead of an individual officer for appraising the performance of the NGOs can be a better solution.

Corporate Organisations too can contribute towards rural development by way of (i) adopting certain villages particularly adjoining their industries, (ii) taking up development projects in rural area, and (iii) channelise their resources for rural development through voluntary organisations. Rich entrepreneurs who are inspired by philanthropy, the field of rural development offers them an excellent opportunity.

4. Adopting Appropriate Technology in Agriculture and Related Fields

4.1. Multiple Cropping Through Irrigation Support

Although there has been improvement in the agricultural produce during last two decades, the overall production has been varying from year to year basis mainly depending upon the quantum and the timely rain. Table 3.10 shows growth rate in foodgrain production from 2000-01 to 2004-05.

Table 3.10: Growth Rate in Foodgrain Production from 2000-01 to 2004-05

Sl. No.	*Foodgrain*	*Production/Growth Rate over Previous Year*				
		2000-01	*2001-02*	*2002-03*	*2003-04*	*2004-05*
1.	Foodgrain production (million tonnes)	195.9	212.9	174.2	212.0	206.4
2.	Foodgrain production (% growth over the previous year)	(–)6.6	8.7	(–)18.2	21.7	(–)2.8

Source: Economic Survey (GOI) from 2001-01 to 2004-05/Statistical Abstract 2003 (GOI).

For achieving higher production as well as better productivity, there is a need to use high yielding variety (HYV) of seeds and chemical fertilizers and irrigation is the prerequisite for the use of HYV seeds and chemical fertilizers. Therefore irrigation facility has become essential to achieve better productivity.

Presently about 35 per cent of the total cropped area in India is having irrigation facility. Irrigated Farming means cultivation for which irrigation facility is available. Perennial source of water provides full-fledged irrigation facility throughout the year. It could be supply of water by canal and installing pumping facility. However in Indian conditions, it is mostly through minor irrigation schemes of various kinds which by itself is an appropriate technology for farming. The construction of small ponds for capturing and storing rainwater serves the purpose of irrigation during off rainy season. Tube wells with small distribution network have also been found to be quite useful.

Five lakh water bodies once helped irrigate nearly four million hectares. This is now down to two million hectares. If these can be restored by repair/renovation, irrigation potential doubles immediately. Water-use efficiency in Indian agriculture is lowest in the world. It is therefore necessary to promote micro-irrigation technology, comprising drip and sprinkler irrigation. So far about 1.2 million hectares have been covered under micro-irrigation

Irrigated farming facilitates multiple cropping and increases yield per ha. During 1999-2000, area sown more than once is 48.51 mha as against the net area sown 141.23 mha. More area can be brought under multiple cropping by tapping irrigation potential available. Taking the various factors into account, the ultimate irrigation potential in India is estimated to be around 139.88 million hectares and as on 31.3.2002, 93.98 mha has been created, with major and medium irrigation projects contributing 37.08 m.ha. and minor irrigation contributing 56.90 m.ha. However looking to the lowering of water table, year by year, the figure of ultimate irrigation potential appears to be overestimation.

Further it is disturbing to note that the net area irrigated has reduced from 56.76 mha in 1999-2000 to 54.68 in 2000-2001. The reduction in the electricity sold to agriculture as a result of load shedding in rural area could be the reason for reduction in the net area irrigated in 2000-01. This highlights the importance of providing uninterrupted power supply to agriculture. The only other reason for reduction in irrigation could be the less quantum of rains in the area of irrigation.

Irrigated farming involves capital expenditure as pumpsets, electricity and sprinkles are required for harnessing the water, apart from use of expensive HYV seeds, chemical fertilizers and pesticides. However these costs are generally offset by the increase in productivity and as such it is advisable to use the appropriate technology as stated above.

4.2. Dry Farming

Nearly 65 per cent of the total cropped area in India is rain dependent. Ways and means have therefore to be found to capture rainwater and retain it as long as possible. One of the ways is to implement water shed management by constructing small ponds for retaining water. The available water can be effectively used by employing the technique of sprinkler and drip irrigation. If the gradient of slope is more, the bunds can be constructed to trap the water.

The other modes to retain the water are: deep ploughing of land before sowing; sowing the crops in ridges, sowing the crops across rather than in the direction of slope; sowing in the form of countours. The spacing between the plants should be more in dry farming in relation to irrigated farming so that the plants can get sufficient nutrient.

The selection of crop should be such that it takes less time for harvesting say 3 to 4 months so that it does not get affected if there is no rain during post-monsoon period. The rotation of crop every year is also beneficial in dry farming. In all these matters, the extension services can play important role in guiding the farmers in dry land farming.

Some of the research laboratories have come out with drought resistant seeds which could sustain low rain conditions. However these have to undergo field trials in the soil conditions as obtained in India and if the trials are successful, these can be utilised by the farmers.

4.3. Organic Farming/Ecology Farming/Vermiculture

Now a days chemical fertilizers and pesticides are being extensively used for the purpose of increasing the yield and quick remedy for the pests. However use of chemical fertilizers and pesticides is found to be harmful in the long run as the produce from these are likely to carry some residues of chemicals. Further, the chemicals can change the natural composition of the soil, create imbalance in the nutritional value of the soil, deteriorate the condition of the soil and contaminate the soil and the water. In view of these likely implications, there is a growing opinion in favour of organic farming.

In this farming, no pesticides or manufactured chemical fertilizers will be used. Instead manure is prepared by mixing cowdung with water, *gomutra* and black jaggery for speedy fermentation. After a week's time the rich manure is released along with the water in the fields. In another process, waste products from household or leftover of the crops after harvesting are stored in a pit for some days, which get converted into the fertilizer called compost fertilizer.

There is a process called Vermiculture in which compost fertilizer is generated by using organic waste and bacteria. The first step is to make vermicastings (beds) on which wet waste is laid. Vermicastings are made by deep burrowing bacteria and covered with leaves and cow dung for 10 days. Wet waste is then spread on this layer expediting its fermentation into a biodegradable product. Rock dust is spread on this to avoid acidic nausea mainly caused by meat and fruits. This area is watered and subsequently more waste is piled on its layers. The compost is ready within a year. Some farmers advocate planting *neem* trees alongside the fields which bring the birds who eat up the insects that threaten to damage the crops. Like chemical-free fertilizers, there are chemical-free pesticides such as *Nimar* etc.

The produce from organic farming is considered to be healthier as it is free from the chemicals. Although there is a greater use of chemical fertilizers and pesticides in developed countries, there is a growing demand of organic farm produce in some sections of people in USA and EU countries so much so that they are ready to help to set up such farms and users are ready to pay extra price for such produce.

4.4. Development of Wasteland

Wasteland in the form of deserts, degraded forests, ravines etc constitutes around 20 per cent (63.85 million ha) of the geographical area. Israel too comprised of vast deserts but they have made spectacular progress in developing the wasteland into excellent agricultural land. India can very well learn from the Israel experience and adopt their methods.

National Wasteland Development Board has been constituted to develop the wasteland. Among other activities, it aims at bringing large area of wasteland under fuel-wood and fodder plantations. The wasteland can also be developed by way of giving ownership rights to poor people through cooperatives and giving tree pattas for growing fruit and other trees. However a care has to taken that the land which is essentially required for the use of community as a whole is not parted with to serve the interests of only few families.

4.5. Combating Degradation of Land

There are many reasons for degradation of soil, the main reasons being drought, water logging, human induced water erosion, desalination and extensive use of chemical fertilizers. In 1999, Govt. estimated that nearly half of the India's 329 million ha of soil could be catagorised as degraded. If degradation of land goes unchecked, the productivity will be affected badly in the long run. It is therefore necessary to institutionalise the efforts of finding ways and means for combating further degradation of land and how the situation can be retrieved.

4.6. Integrated Pest Control Programme

Farmers usually spray pesticides individually in their farms. Some of the farmers do not use pesticides, either because they can not afford or because they are not aware about the importance of the pesticides. In such situations, the pests which remain active in the farms where pesticides have not been used, can after a while damage the crops in the farms where the pesticides were earlier used. Conversely, the remnant of the pests from the farms where pesticides have been used can damage the crops in the farms where pesticides have not been used at all. In the circumstances the most effective way of using pesticides could be if they undertake a collective programme of using the pesticides so that the whole contagious area is cleared of the pests.

4.7. Cooperative Farming

Land Holdings are getting fragmented from generation to generation on account of divisions in the families. As a result, there is a steady increase in the number of small and marginal farmers. Table 3.11 shows classification of operational holdings and area operated in 1985-86 and 1990-91.

Table 3.11: Classification of Operational Holdings and Area Operated in 1985-86 and 1990-91

Category of Holdings		*Operational Holdings*		*Area Operated*	
		(million)	*%*	*(million Ha)*	*%*
Marginal		62.11*	59.0	24.62	14.9
(below 1 ha)		(56.15)**	(57.8)	(22.04)	(13.4)
Small		19.96	19.0	28.70	17.3
(1 to 2 ha)		(17.92)	(18.5)	(25.71)	(15.6)
Semi-medium		13.91	13.2	38.35	23.2
(2 to 4 ha)		(13.25)	(13.6)	(36.67)	(22.3)
Medium		7.63	7.2	45.05	27.2
(4 to 10 ha)		(7.92)	(8.1)	(47.14)	(28.6)
Large		1.67	1.6	28.90	17.4
(10 ha & above		(1.92)	(2.0)	(33.0)	(20.1)
Total	(1990-91)	105.28	100	165.62	100
	(1985-86)	(97.16)	(100)	(164.56)	(100)

* Figures pertain to 1990-91

** Figures in parenthesis pertain to 1985-86

Source: NIRD Rural Development Statistics, 2002-03.

The following trends are seen from the aforesaid table:

1. Land is getting fragmented from year to year basis.
2. There is reduction in the large and medium holdings and increase in marginal, small and semi-medium holdings.

3. Marginal (59) and small holdings (19) together constitute 78 per cent or say 80 per cent of holdings.
4. Marginal and small category of farmers with 80 per cent holdings operate 32 per cent of land area and medium and large farmers with 20 per cent holdings operate 68 per cent of land area.

Arranging separate resources for cultivation by each farmer becomes difficult and uneconomic. In this context, it is better if such farmers form a cooperative, pull the resources together and do the farming jointly. In this case the members will continue to be the land owners but the cultivation will be done by the cooperative. Any services lent by the members will be paid by the cooperative. The principle of cooperation can be extended to marketing as well. Recognising the benefits of cooperation, Gandhiji said, "I firmly believe that we shall not derive the full benefits of agriculture until we take to cooperative farming cooperative farming of my conception would change the face of the land and banish poverty."

Notwithstanding apparent benefits involved, there is a lukewarm response to cooperative farming as it is seen that only about 2 per cent of the cultivators have formed cooperative farming societies covering about 0.2 per cent of the total cropped area. Lack of local initiative and absence of leadership in the farming community are the main reasons for the poor response. However as time passes there does not seem to be any choice for larger section of fragmented landowners but to join together to make the cultivation profitable or in some cases even sustainable. In this context the State can play proactive role in finding ways and means to popularise the concept of cooperative farming.

There is also a suggestion that contract farming if made as a part of cooperative farming may help the situation, as this will strengthen tie-up between cooperatives and private sector. However this concept is yet to be tried and tested. The main point in this concept arises what will happen to the agricultural labour presently engaged in the new set up?

4.8. Development of Khadi and Village Industries (KVI) and Village Small Industries (VSI)

The KVI scheme is intended to promote rural industrialisation. In order to help capacity building and technology up gradation, it is planned to upgrade the Wardha Institute started by Gandhiji in 1935 as a national institute to be called Mahatma Gandhi Institute for Rural Industrialisation. However the main problem for the growth of these industries is the assured market. For example, most of the people prefer mill made cotton instead of khadi due to fineness of mill made cotton and its maintenance is easy. However khadi is most suitable for Indian climatic conditions. Besides, it gives employment to rural population. In this context it is a right step that Indian Railway will use only khadi for its linen so long as Khadi and Village Industries Commission (KVIC) can fulfill its requirement. Similar support should be extended to other cottage industries by various organisations.

In order to market the products of Small Industries, the quality of the product, packaging and cost have to be competitive in the National/International market and for this, they need infrastructure such as quality control laboratories, cold chains/storages, packaging units etc. There is also a need for constant technological up-gradation of processes and equipments so as to keep pace with the development in our country as well as other countries. Govt. has to provide these inputs for desired progress of this sector, which is essential for the rural development. Also KVI are facing financial crunch and therefore there is a need to establish separate financial agency for providing to them timely required working capital.

4.9. Dairy Development

Presently India ranks first in milk production in the world. The production in the country has been assessed as 88.1 million tones during 2003-04. India's share in the total world export of milk and milk products is about 0.24 per cent. There is a great demand for milk and milk products both in rural as well as urban area. Due to constraints like non-availability of space and pastures and problem of pollution, it is difficult to develop dairies in urban area. Since the space and pastures are available in rural area, dairy development can be conveniently undertaken there. The dairy development is a flourishing business and also offers good employment opportunities outside agriculture. It has already undergone large expansion. However the success achieved in Operation Flood I & II in Gujarat and White Revolution in Punjab and Haryana should be made known to the rural people in other parts of the country especially Bihar,

Goa, Chattisgarh and Jharkhand apart from North Eastern states which are particularly deficient in milk production at present as in these states the per capita production of milk is less than half the national average.

Our country has the best buffaloes in the world. This black gold has been a neglected animal over the years. Their genetic potential needs to be improved. A new joint venture of India Immunologicals ltd. And Co-operative Resources International (CRI), USA has been formed to provide breeding and advisory services to the milk producers and farmers. Artificial insemination has been cited as one of the methods to improve the yield of milk.

The business of milk involves preservation of milk necessitating cold storage and transportation of products to urban centers. Therefore it can be more profitable if it is done on a larger scale by way of forming cooperatives and pulling resources together. National Dairy Development Project can play a crucial role in giving thrust to organise the cooperatives in the rural area. In this respect in the arena of cooperative system Amul dairy at Anand in Gujarat is a great success story which has been acclaimed world over. It is worth emulating at other places too in the country.

4.10. Krishi Saptak Programme in Maharashtra-Recommendations of Dr. M.S. Swaminathan Committee

Looking to the challenges ahead during 25 years hence for the development in the field of agriculture, Maharashtra State constituted a Committee under the Chairmanship of Dr. M.S. Swaminathan, noted expert in agriculture. The Committee submitted its report on 31st December 2002. From among various recommendations of the Committee, the following seven main recommendations have been agreed to be implemented by the Maharashtra state under Krishi Saptak Programme from the year 2003-04.

- (i) Extension of crop insurance for dry farming in stages
- (ii) Conducting on large scale, programmes like organic farming for increasing productivity of dry farming, integrated pest control, use of vermicultural compost fertilizer, production of organic fertilizers and pesticides etc. and giving guidance literature on testing aspect to farmers.
- (iii) To take up programmes of dissemination of information and

expertise about agriculture and fruit production so as to give guidance to farmers about agriculture and marketing aspect; and establishment of Virtual University for transfer of technology

(iv) To establish various laboratories for examining grade of agricultural produce and creating number of information centres.

(v) To promote participation of women in agriculture related programmes.

(vi) To promote creation of self-employment opportunities in agriculture and related areas in a big way.

(vii) To constitute under the Chairmanship of Agriculture Minister, a Agriculture Parishad comprising progressive farmers, experts, scientists etc who are associated with the State's agriculture and related fields.

5. Institutional Development

5.1. National Water Policy

Water is the lifeline of civilization. It is told that the biggest crisis that the world will face in the 21st century will be the crisis of water and the wars in future will be fought on water. Water is a prime natural resource, a basic human need and a precious national asset. Optimum development and efficient utilization of water resources, therefore, assumes great significance. The Ministry of Water Resources lays down policies and programmes and development and regulation of the country's water resources.

Water Resources Potential: The average water availability of the country is assessed as 1869 bullion cubic metres (BCM). Of this, total utilizable water resource is assessed as 1122 BCM, comprising surface water 690 BCM and ground water as 432 BCM. The per capita availability of water at National level has reduced from about 5, 177 m^3 (cubic metres) in 1951 to the estimated level of 1,820 m^3 in 2001 with variation in water availability in different water basins. Looking to the importance of the water in India, National Water Policy (Draft)-2002 calls for integration of water resources development and management for optimal and sustainable utilisation. But it excludes inter-state water disputes, which are usually of very intricate nature.

National Water Policy underlines the idea, recognising water as a national resource and access to save adequate and affordable water as a basic need of every citizen with the right to drinking water as a priority to other alternative use. Depletion of groundwater resources, on which millions of rural families depend for their drinking water needs as well as irrigation, continues unabated. By 2050, India will have to support a population of around 1.6 billion with increased requirements of potable water, food (450 million tones) production and industrial use. The situation has been made worse with growing pollution and inefficient use of surface water. The policy therefore lays down the principle 'polluter must pay for the pollution of the water resources'. There is a need for rainwater harvesting, conservation of local water resources, equitable and efficient allocation of irrigation water and conservation of groundwater along with treatment of wastewater.

Regulation of Ground Water Use: Faced with depleting ground water levels, the Centre has advised the states to enact Bills to regulate usage of hand-pumps and concentrate more on rainwater harvesting and conservation projects. The depletion of ground water has affected the quality of water through fluoride and arsenic contamination and the Centre has circulated a draft Bill for enactment by the State Assemblies. The contamination and brackishness have already forced many State governments to abandon low-cost hand-pumps and turn to costly piped water supply schemes.

Further while generally there are restrictions how deep a borewell could be (maximum 60 metres) for drinking purposes, there is no such limit for borewells dug up for irrigation. Such a restriction is needed to maintain groundwater levels.

5.2. Interlinking of Rivers

Some parts of the country receive much more than normal rainfall leading to floods thus causing huge damage to life and property, especially in the Ganga and Bramhaputra rivers. On the other hand large chunk of the peninsular India receive less than normal rainfall, leading to droughts. Water resources in Gaga and Bramhaputra basin make up 60 per cent of India's total water resources. The proposal of interlinking of rivers is borne out of the concept that it will solve the aforesaid problem of floods and the droughts. The Constitution stipulates provisions for the Centre to play the role of a trustee in matters related to water. Entry 56 in the Union List calls for "regulation and development of inter-state rivers and river valleys to the extent to which such regulation and development under the control of the Union is declared by Parliament by law to be expedient in the public interest". Interlinking of Rivers Plan aims at retrieving floodwater causing havoc and going waste. The proposal of interlinking of rivers envisages linking 37 rivers through 31 links and 9000 km of canals. This also involves creating a network of water storage reservoirs across the country.

The rough estimate indicates that the project would cost Rs. 5,60,000 crores to obtain 175 billion cubic metres (BCM) from interlinking of rivers. However apart from financial resources, some experts have strong reservations about the very concept of interlinking of rivers. They feel such high technological interventions are unnatural, eco-unfriendly, may actually be counterproductive and also lead to large-scale displacement of affected families. The Environment Impact Assessment, Feasibility and Techno-Economic Studies have to precede starting actual implementation of the project of this magnitude. It is pertinent to note that the Consultants' costs for the aforesaid studies are going to be huge. Further, some of the states are reluctant to allow even preliminary hydrological studies to be conducted in their states especially on inter-state rivers, under the presumption that that this may affect their present share of water resources. Already there have been simmering disputes over Narmada and Cauveri water and now added to the list is Sutlej Yamuna Link. Rivers are the lifeline of the people. In the Indian context, they are not merely channels of water but embodiments of a living culture. Constitutional provision alone may not solve the problem unless the local population is convinced about the overall interest and their own interests are fully taken care of. The broad feasibility report is expected to be ready in 2005-06. In view of the larger issues involved and to avoid wastage of scarce financial resources, there is a need in the first instance to evolve broader consensus on the proposal before taking up actual work on Inter Linking of Rivers at ground level.

In any case priority will have to be given to interlinking of south-bound rivers since some rivers in the north have international ramifications.

5.3. Artificial Rain by Cloud Seeding

Agriculture in India is by and large still dependent on rain. There are often instances of long gap after the rain stops and the crops start getting withered away. Sometimes there is a long gap just after the sowing of seeds is completed and the farmers have to sow the seeds again. This involves not only extra expenditure but also the crop looses the rhythm. To ameliorate such a situation, Maharashtra State launched a "Project Varsha" to receive artificial rain on experimental basis. Radar systems have been installed at Shegaon and Baramati. The radar is installed atop a high-rise building, which helps in tracing the clouds. The information is then passed on to the pilot of the piper plane, which undertakes seeding operation by spraying Silver Iodide, Sodium Iodide and Sodium Chloride on rain clouds. It then starts raining within a few minutes. The result depends upon whether the clouds are weak or strong and the quantum of water being carried by them. The above method of receiving artificial rain is not new. It is being tried against the background of conditions of consecutive droughts in Maharashtra. The results of the experiment have to be tested on techno-economic basis over reasonable period of time before it is implemented on wider scale.

5.4. Land Reforms-Digitisation of Land Records

There are glaring shortcomings in the implementation of land reforms as mentioned earlier. Out of 7.35 million acres declared as surplus, possession of only 6.48 million acres was taken over and 5.30 million acres were actually distributed. There is a need to take over the remaining land and distribute to landless and marginal/small farmers.

On account of Land Ceiling Act and divisions in the joint families from generation to generation, land holding per family is becoming smaller and smaller thereby increasing the number of small and marginal farmers. Due to high input cost, a small piece of land is not viable for cultivation unless it has irrigation facility. Against this background, consolidation of holdings by means of pooling of land by a group of villagers is one of the alternatives. It is like forming a cooperative by the small and marginal farmers for cultivation as well as marketing. This has been tried at number of places with varying degree of success.

The land records are maintained by the *Patwaris*. Since the records are maintained manually there is a scope of manipulation of the records and there have been number of disputes on this account. Due to this difficulty, in some parts of the country the land records have been digitised on computers at Taluk level and this duel system has been found to be quite successful. This needs to be followed in all parts of the country. Further the updating of land records is a continuous process. For doing this work, the revenue administration has to be revamped and it has to be provided with necessary infrastructure/modern survey equipments like Global Positioning System (GPS), EDM, Total Stations, Theodolites, Work Stations etc. to take up the fieldwork.

5.5. Restructuring of Rural Banking System

Large institutional framework has been established in the rural area for providing credit to agriculturists, artisans and for starting cottage industries. There are more than 33,000 branches of commercial banks, 14,500 branches of regional rural banks and 1,35,000 rural credit co-operatives.

The implementation of the latest policy on agricultural credit, announced by GOI on 18th June 2004, is entrusted to public sector and private sector banks, Regional Rural Banks (RRBs) and the Cooperative Banks. It is proposed hereafter to double the flow of agricultural credit in three years. In 2004-05 itself, it was planned to increase the gross credit to agriculture to Rs. 1,05,000 crores, 40 per cent up from Rs. 70,810 crores disbursed in 2003-04. As against these targets, presently the health of cooperative and regional rural banks, which together with public and private banks have to play active role in this venture, is pretty bad and they need complete restructuring.

5.5.1 Cooperative Banks

In past the cooperative banks have definitely helped agriculture sector in providing credit without an element of overexploitation as happens in the case of private moneylenders. However the sector has reached the level of plateau and no growth is taking place in terms of total volume of credit. On the other hand there is decline in financial investment in agriculture sector as a percentage of the total investment in the whole economy.

The Board comprising Chairman and the Directors manages the state cooperative banks. However the final administrative authority rests with the State Govt. On the other hand functionally they are under Reserve Bank of India in as much as these banks are supposed to run as per the guidelines of the RBI. RBI also periodically inspects working of these banks and suggests actions to be taken for any deviations. This has lead to a situation of dual control of these banks. The State and the management of the banks tend towards accommodating the needs of the farmers even in the cases in which the criteria for getting the credit is not fully met. A large number of farmers got benefited from this approach. However this flexible approach, which is mixed with genuine social concern in some cases and vested interests in other, has led to large amount of credit remaining unrealised. Some of the cooperative banks also invested their money in Govt. securities through middlemen and some of these middlemen did not procure and deliver the securities to the banks. These cooperative banks not only lost money but the sector as a whole got a bad name in spite of the fact that these banks did provide service to the community in its difficult days.

Across the country the recovery position is of course unsatisfactory in as much as the overdues amount to almost 20 per cent of the total loan outstanding. In this context, the practice of waiving off the old outstanding loans as a populist measure followed sometimes just before elections also deters lawabiding farmers to repay the loans and emboldens the law breakers. Instead, the farmers can be offered reduction in the interest rates and easy installments for repayment. As it is while NABARD refinances banks at the interest rate of about 6 percent for farm credit, the actual interest rate paid by farmers more than doubles, which is now being reduced to one and half times. Another aspect is that although the objective of the cooperative organisation was to serve all sections of the farmers especially small and marginal farmers who need the credit most, unfortunately the benefit mostly did not reach this section of people. The main reason for this is that the managements of the societies are in the hands of bigger farmers and they are not very sensitive to the requirements of the small and marginal farmers. Further the Chairman and Board members manage to continue in the same positions for any number of years thus precluding the chances of implementing any innovative systems that may be required to keep the organisation abreast of the time to face the challenges from the contemporary competitive environment. A time limit of 3 to 5 years on the tenure for management positions with the rider that they would not be allowed to contest the society's elections again may serve the purpose.

Gandhiji was a strong proponent of the co-operative concept and in particular about the necessity of its application to agricultural sector. All the same he foresaw, even at the conceptual stage, the danger the co-operative movement is likely to face when he said, "The co-operative movement will be a blessing to India to the extent that it is a moral movement strictly directed by men fired with religious fervour. It follows, therefore, that co-operation should be confined to men wishing to be morally right".

Unfortunately there are few persons who fit into the concept of Gandhiji. One such person who made a mark in the Indian history of co-operative movement is Verghese Kurien. He was General Manager when Lal Bahadur Shastriji, the then Prime Minister visited AMUL (Anand Milk Union Ltd.) in Gujarat in October 1964. The Prime Minster was so impressed by the AMUL model of dairy development that he asked Kurien to prepare a scheme for replicating the model throughout the country. It led to establishment of National Dairy Development Board (NDDB) with its headquarter at Anand. The PM appointed Kurien as Chairman of NDDB, a position he held from 1965 till November 1998. Thereafter he was made its Chairman *Emeritus*. Presently he is Chairman, Institute of Rural Management, Anand. He has been instrumental in conceiving the project named, "Operation Flood-creating virtual flood of rurally produced milk in the country" and attaining the very success of it. The work of Kurien and under his leadership, the success of AMUL co-operative has been acclaimed by eminent world bodies.

Unless cooperative banks are healthy and credit worthy, it would not be possible to reach credit to every farmer in need of credit. Presently the situation of cooperative banking system is pretty bad. Six State Central Cooperative Banks and 140 District Central Cooperative Banks do not comply with Section 11 of the Banking Regulations Act, 1949. They also have difficulty in accessing refinances for agricultural credit. Similar is the case of land development banks. In order to find a durable solution, tasks force headed by A. Vaidyanathan was constituted in 2004 to examine the reforms required in the cooperative banking system including the regulatory regime. The Task Force submitted its report to GOI on 5th January 2005. The major recommendations are infusion of fresh capital in cooperative banks as was done in PSU banks in 1990s; resolution of the issue of 'duality of control' by the Reserve Bank of India and State Cooperative Acts; Reducing multiplicity of layers and widespread inefficiencies that has eroded the profitability of cooperatives besides substantially raising the cost of borrowing.

To act upon the recommendations of the Report, the GOI has decided that if the states are ready to accept the conditionalities for bringing about the legal, administrative and structural changes and to participate in the scheme and share the financial liabilities, financial assistance would be released linked to the conditionalities. The assistance would cover invoked but unpaid guarantees given by the State Governments, increasing the state capital to a specified minimum level, retiring Government share capital and technical assistance.

5.5.2 Regional Rural Banks (RRB)

Each Regional Rural Bank has a sponsor bank. RRBs are required to increase their loan target by 40 per cent. Sponsor bank is being made responsible for the RRBs under its control. There are 198 RRBs most of which are ailing. RRBs that adopt a new governance standard and that abide by the prudential regulations will qualify for receiving funds from the Govt. for their restructuring.

The failure of banks largely happen on account of non-evaluation of credit request proposals from the point of view of economic viability, repayment capacity, infrastructure support to and functional linkages of the purpose of loan with other related activities.

5.6. (Rural) People Friendly Credit System

At one end of the paradigm small farmers and weaker sections of the society could not derive much benefit from various schemes and at the other end, overdue position of the banks is showing little improvement The position can improve by making extension services as part of the credit so as to enable small farmers to get maximum return from small family holdings. The staff of the Block Development Offices/Agriculture officers and special cells in the banks to deal exclusively with the matter of providing credit to the farmers can greatly help the situation. Further, banks should assess credit worthiness of farmers on a case to case basis so that their entire land need not be mortgaged.

Time is the essence in agricultural operations. If the credit is not delivered at right time, it is of no use. Further, credit should be adequate enough to fully meet the requirement of all the operations of the farming till the produce is sold in the market.

The interest structure of the cooperative banks has not been very favourable to the poor borrowers. Typically if we add interest charges charged by each tier i.e. NABARD-5 per cent, SCB-3 per cent, DCCB-2 per cent, PACS-2 per cent, finally the borrower has to pay interest rate of ranging from 12-16 per cent. Reducing inter bank rate, the interest rate to the borrower is now proposed to be reduced from 12-16 per cent to 9 per cent.

There should be a differential rate of interest for the rural poor. As it is, some commercial banks have announced interest rates as low as 7.5 to 8 per cent for building houses in the urban area-It is unfortunate to see the trend that the commercial banks prefer consumer and corporate borrowers at the cost of individual borrowers mainly in the small scale and farming sector by way of charging less interest to former in relation to latter. This is reminiscent of the past era when there was caste bias in charging the rate of interest by the private lenders. The scheme of lending should be on the basis of individual borrowers' track record rather than on attributable basis. This way the banks can offer lower interest rates to honest borrowers.

Since presently multiple credit schemes are available and the farmers have to go from desk to desk it would help if there is a single window system in true sense. Loan melas have also been found to be quite helpful in getting the credit quickly. Such schemes will however be more successful if the same are not politicised.

On the part of small farmers and weaker sections of the society, there is a need to adopt a concept of family budgeting and personal savings, whatever little it may be, for taking care of the expenses of the whole year and also social security for future. One of the a!reas could be, observing austerity in family functions such as marriage ceremonies etc. NGOs can play important role in creating awareness among the rural population about such aspects.

5.7. Integrated Kisan Credit Card

There used to be every year hassles for the farmers for getting the loan from the lending banks. The Kisan Credit Card will not only do away with these hassles but the holder of the card will have readily available with him all the information about the loan he has availed. It is like a bank pass-book. This will definitely improve comfort level for the farmer. Presently there is a loan limit of Rs 50,000 under credit card scheme. However there are numerous reports that the farmer has to shelve out (unofficially) handsome money to get the card issued. There is a need to ensure that the issue of the credit card is made corruption free. Further there can be integration of important services like health insurance and crop insurance in the same card, so that the farmer has a complete record with him about all important matters concerning family welfare as well as agriculture.

5.8. Effective Extension Services

There is so much influence of the traditional way of doing agriculture on the rural people that it requires strong persuasion to make them adopt new varieties of crops as well as improved methods of doing agriculture. The role of extension services start right from the stage of assistance for soil testing to find out the quality of soil and the type of crop which can be most suitable for that land. The testing also throws light on the deficiency of nutrients in the soil.

Farmers have to be convinced about the advantages that they will reap by adopting new crops and improved methods of farming. For example in the case of small and marginal farmers, if there is a facility of even small irrigation, cultivation of fruits and vegetables can be quite advantageous. The extension services incorporate all aspects of agricultural operations starting with preparing the field for sowing to marketing the produce. These include selection of proper variety of the crop, proper ploughing, using high yielding variety seeds, sowing in time, selecting fertilizers to provide correct nutrients, irrigating the crops wherever possible, removing the weeds in time, observing the pests and spaying pesticides that are effective for a particular situation, harvesting the crops in time and preserving them by proper storage before marketing. In all these operations, not only the process is important, but also the timing To make extension services effective, the extension staff have to physically demonstrate these processes in the sample fields so that the farmers get convinced about the benefits as well as they get familiar with the processes so that they start implementing these measures, shedding their usual reservations.

Presently Village Development Officer (VDO) or the Gramsevak are posted in the villages to assist the villagers including farmers in various development matters. In bigger village VDO is posted and in smaller one gramsevak is posted. The VDO/Gramsevak reports to Block Development Officer. As regards extension services these are provided through village extension worker (VEW)/Agriculture Supervisor (AS) who reports to Agriculture Officer at Taluk level and District Agriculture Officer at the district level. Each extension worker/AS looks after nearly 1500-3000 farmers. Looking to the scope of services to be provided, this number is too large for him to handle. In China the area allotted to each extension worker is one-fourth of that in India, Thus there is a need to provide more number of extension extension workers particularly in the context that the adult literacy rate in rural area in India is less than that in China. Further there is a need to prescribe proper qualification (B.Sc. agriculture) for the appointment of extension worker and also provide adequate training to him so as to make his services more effective.

Presently extension agency brings to the farmers the knowledge of developments that have taken place as a result of R&D effort in agriculture. When the new practices are implemented, farmers may come across various difficulties. The extension agency has therefore to give inputs to the other end i.e. research and development centers about the benefit or the practical problems encountered in implementing the modifications. This feedback guides the R&D centers to proceed with further research or offer suitable solutions for the problems being faced in implementation of the innovations suggested by them. However somehow this mechanism of feedback is not working as desired and there is a gap in the dissemination of information between the farmers and the R&D centres. It is therefore necessary to exchange on regular basis the information between the farmers and the R&D centers about the implementation of new practices, problems encountered in their implementation and solutions thereof. Acknowledging the importance of the extension in agriculture, the centrally sponsored scheme – Support to State Extension Programmes for Extension Reforms – to provide need-based support to states for undertaking extension reforms is underway.

5.9. Higher Productivity through Application of R&D/ Technological Advances in Agriculture

During last two decades there is an improvement in quantitywise (tonnage of) production of crops in India. However the yield per acre is low compared to developed countries or even china. Table 3.12 shows average yield of important crops during 2002-03 in India.

Table 3.12: Average Yield of Important Crops During 2002-03.

(Quintals per Hectare)

Crops	*Yield*
Rice	18.0
Wheat	26.2
Pulses	5.6
Total Foodgrains	15.6
Total Oilseeds	7.1
Cotton	1.9
Sugarcane	645.6

Source: Statistical Abstract 2003 (GOI Publication)

It is pertinent to note the extent of improvement in productivity of agriculture *vis-à-vis* manufacturing and services. Average labour productivity is measured by the sector gross state domestic product divided by number of workers employed in the sector. The productivity index of agriculture increased from 1 to 1.5 between the early 70s and the late 90s, whereas the productivity of manufacturing increased from 2 to 4.5 and that of services has increased from 2 to 5.5 over the same period. Within the services sector, productivity per unit of labour is the highest (30 as against 1.5 in agriculture) in the minuscule organised sector. While the productivity of agriculture has grown gradually over the years, that of manufacturing and services has increased speedily. In spite of aforesaid increase, the average productivity in agriculture in our country is low even compared to China. For example, wheat yield in our country is 2.62 tonnes per hectare (TPHA), which is a way below the 3.9 tonnes per hectare of China. Geographically although China's area is three times that of India, its cropped area is almost the same as that of India. Even then China produces almost one and half times more foodgrains than India. It is a matter of great concern that the productivity of land now is either stagnating or declining. The main reasons for low productivity in India are too much fragmentation of land holdings, heavy dependence on rain, low growth of irrigation, lack of proper infrastructure and absence of mechanisation in the farming as most of the agricultural operations in India are done manually. The farmers are also not in a position to make higher investments in agriculture due to their low economic status and not being sure of the returns.

In the above context, Research and Development can play a crucial role in bringing all round development in agriculture. Indian Agriculture Research Institute, Agriculture Universities and Agriculture Colleges are doing extensive research in upgrading farm technologies and developing improved variety of seeds and within the resources available to them, they are doing commendable work in these fields. Private participation in R&D can further reinforce the efforts of these institutions. However the investment in agricultural research in India is abysmally low at 0.26 per cent as against the investment of 2.4 per cent in developed countries. Further in order to enable the agriculture to derive full benefit of the laboratory facilities and the research work being conducted, it is essential to adopt the "lab to land" programme so that the orientation of research work is directed towards practical use in agriculture.

5.10. R&D in Agriculture Engineering Tools

The engineering tools have to be designed to meet specific needs. If the land size is big, large capacity tractor will be better, whereas in small size land, a small size tractor will only be useful. Similarly implements like seeding ploughs, weeding instruments, threshers etc are undergoing constant redesigning to suit the prevailing conditions of soil, topography and type of crop. Improvements in engineering implements play a vital role in increasing the productivity and the developed countries are using mechanisation in agriculture extensively. The land conditions in developed countries are of course different in the sense that their farm sizes are bigger and extensive mechanisation is feasible.

5.11. R&D in Creating High Yielding Variety (HYV) of Seeds

Research & Development in the direction of increasing productivity lead to identifying high yielding variety (HYV) of seeds particularly in respect of jowar, wheat, paddy, cotton etc. The output from these seeds has increased manifold. However use of these seeds require many other inputs like intermittent watering, fertilizers, pesticides etc. HYV has led to a whole new concept of farming. Since this system is investment oriented, farmers need more credit facility for purchasing whole set of inputs. If the supply chain is properly maintained, the farmers can look for better days ahead.

In the context of seeds being developed by the farmers, a new Seed Bill introduced in Parliament will replace the Seed Act of 1966. The Protection of Plant Variety and Farmers' Rights Act, 2001 (PPVFR) which is in place requires the declaration of the origin of the variety with pedigree details, which the Seed Bill does not. The bill therefore requires to be reviewed if by provisions like this one, the interests of the farmers will in anyway suffer vis-à-vis seed industry.

5.12. R&D in Bio-Technology/Genetic Engineering/Genetically Modified (GM) Crops

The yield with high intensity agriculture i.e. using chemicals-fertilizers-pesticides and mechanisation has reached its physical limits. To remain competitive, research is constantly going on in the area of developing the seeds and plants that can give better productivity and nutritive value. Cross breeding can produce improved varieties but it is difficult to introduce new characteristics within limited gene pool, hence the need for Genetically Modified technology. GM could enhance agricultural productivity by introduction of genes from same or other species. It is claimed to reduce reliance on agrochemicals by producing disease-pest resistant varieties and decrease water requirement by adoption to drought. Genetic experiments are being conducted on cotton, maize, mustard, sugarcane, sorghum, pigeonpea, chickpea, rice, tomato, brinjal, potato, banana, papaya, cauliflower, oilseeds, castor, soyabean, and medical plants. Experiments are also conducted on several species of fish. Presently the companies which are in the forefront of the research in GM technology, are Monsanto and Cargill, both multinationals. Genetic Engineering Approval Committee (GEAC) under Ministry of Environment & Forests is the approving authority for the results. Bt cotton is the first transgenic crop in our country. Although field trials were conducted on Bt cotton and GEAC approved the results, bio-safety concerns are often raised as to its suitability for Indian conditions. Besides looking to the aspect of advantages of productivity and nutritive value of GM food, there can be no two opinions about fully investigating the negative and harmful impact of any of the GM crops and minimise them to acceptable level before putting them into commercial use. However an overall assessment so far indicates that the benefits from the use of GM plants are substantial. The adoption of Bt cottonseeds by the farmers in the country has been rapid. From 62,000 acres in the year 2002, the area under coverage in 2004 is about one million acres, also indicating "significant agronomic benefits leading to better returns". Trade-Related Intellectual Property Rights (TRIPS) allows patents on genes and cell lines. However, only giant companies are likely to enter into this field in a big way because they will only have necessary resources to undertake extensive and costly research into this highly specialized field,

During the last decade foodgrain production in India has registered a compounded annual growth rate of 3 per cent. If the country has to maintain this growth rate and remain competitive in the international market, we should not close our minds for accepting any new technology, provided it suits us. After all there were serious reservations even at the time of using high yielding variety (HYV) of seeds for the first time, but now these are the only seeds mostly being used.

In order to quicken the flow of biotechnology to crops, there is a need to frame National Bio-technology policy as well as evolve a Single Window Regulatory Body to speed up the approval or rejection of new biotechnologies in order to bring additional choices for farmers.

5.13. Mechanism for Fixation of Electricity Tariff/Should There be Free Supply of Power to Agriculture?

Presently in most of the states less tariff is being charged for the electricity supplied to pumpsets and some of the states have declared free power supply to pumpsets. However due to poor financial health of SEBs, pressure is mounting for increasing tariff in the electricity supply to agriculture. Industrialists' Associations are also voicing this point in various fora a as they feel that part of the burden due to less tariff to agriculture is getting passed on to them. Table 3.13 shows the category wise consumption of electricity in 2002-03. The consumption of electricity in agriculture amounts to 24.88 per cent of the total consumption.

Table 3.13: Categorywise Consumption of Electricity (Utilities only) in 2002-03

Category of Consumer	*Units Sold (Gwh)*	*Per cent of Total Units Sold*
Domestic	83355	24.54
Commercial	25437	7.49
Industrial	114959	33.85
Agriculture	84486	24.88
Others	31361	9.24
Total	339598	100.00

Source: Statistical Abstract–India 2003 (GOI)

The electricity sold to agriculture is however decreasing from year to year. Table 3.14 shows the trend of electricity sold to agriculture from 1998-99 to 2002-03.

Table 3.14: Share of Agriculture as a Percentage of Total Electricity Sold from 1998-99 to 2002-03

Year	*Total Units Sold in the Country (Gwh)*	*Units Sold to Agriculture (Gwh)*	*Electricity Sold to Agriculture as a % of Total Units Sold*
1998-99	309,734	97,195	31.38
1999-2000	312,841	90,934	29.07
2000-01	316,600	84,729	26.76
2001-02	322,459	81,673	25.33
2002-03	339,598	84,486	24.88

Source: Statistical Abstract-India 2003 (GOI)

It is seen that there is a steady decline in the units sold to agriculture in the succeeding year with reference to preceding year, both in terms of absolute number of units as well as percentage of total units sold, with the only exception of the year 2002-03 in which units sold to agriculture are more than the preceding year 2001-02. Incidentally there was drought during 202-03 that perhaps explains the reason for increased supply to agriculture.

In this context it is to be noted that barring a few big landlords, small and marginal farmers who constitute more than 80 per cent of the farmers are not getting each year enough income even for their subsistence. The question arises as to how they will be able to pay high charges for electricity. In developed countries substantial subsidies are given to agriculture although farmers' conditions are not so bad. Agriculturists in India need subsidy for their bare survival. However, simultaneously, the interests of the electricity supplier have to be protected. Already the financial crisis of the State Electricity Boards has impacted the agriculture sector, reducing their ability to undertake required investments, meet rising local demand and maintain day-to-day operations. The burden of demand subsidy should therefore be borne by the Govt. One of the ways to move away from subsidising power is to provide capital subsidy to the supplier for improving access in rural areas. The other proposition could be to facilitate new entry for any innovator-supplier for cost effective electricity supply.

The tariff to be charged from the agriculturists should be worked out by an expert committee which should take into account interest of the electricity supplier as well as the agriculturists. There should however be no free supply of power to agriculture as it amounts to great loss to the electricity supplier. In Maharashtra alone, it has been to the tune of Rs. 1500 Crore per year. Unable to sustain the financial burden, the scheme of free power has been withdrawn by Maharashtra state within one and half year of its introduction. Apart from loss of revenue, such schemes will lead to dispensing of meters in which case it will not be known as to how much power is being supplied to agriculture. Free power will mean there will be no deterrence to willful farmers for non-judicious use of power which may result in excessive drawal and wastage of power. For this very reason, the tariff to be charged to agriculture should be on consumption basis and not flat rate.

Populist measures such as free power announced on the eve of elections is in fact working against the long-term interests of the farmers. Heavy loss of revenue to the supplier is deterring the supplier to provide electricity to agriculture which fact is evident from the steady decline in supply to it from year to year basis as shown in the above table rather than going round about the way. Some of the electricity boards are toying with the idea of changing over 3-phase supply to single-phase supply in rural area. Such a scheme though may be all right for lighting purpose, may result in restricted hours of supply to agricultural pumpsets, besides extra investment on the part of supplier and technical difficulties involved in operation of the system on a large scale. Andhra Pradesh and Karnataka have already implemented the scheme in certain areas and Maharashtra State Electricity Board has taken up this work on pilot basis. In the complex situation like this, application of reasonably subsidized tariff to agriculture can only be the logical best solution.

Some of the state Governments unable to bear the financial burden are withdrawing the free power benefit only a few months years after its announcement. Andhra Pradesh which announced free power to farmers about a year earlier has modified the policy on 31.01.2005. The new policy to be effective from 1st April 2005 is reported to exclude rich farmers from the purview of free power and restrict the benefit to farmers in dry land areas and small and marginal farmers in wetland areas.

5.14. Ushering in Quality Revolution

The quality of the final produce is very important for the farmer as the price of the produce in the local market is directly related with the quality of the produce. It has not only to be of good standard but it has also to stand the stiff competition from other countries. Further the quality of the produce has to confirm the food regulations of the importing countries. In this respect, the regulations in USA and European Union are quite stiff. Since quality of ultimate product depends upon each agricultural operation, the farmers have to be conscious about it right from the stage of preparation of land till the produce is safely received at the purchaser's end.

To check quality of soil, produce or any other matter, there is a need to provide easy testing facilities to the farmers and in this context the 'Lab to Land' Programme assumes great importance. Similarly the cold chains for preservation of perishable produce and container facilities for transportation are necessary.

5.15. Marketing System with Easy Accessibility

In order to get better price for the produce, the role of middlemen should be avoided as far as possible. Agriculture Produce Marketing Committee (APMC), a cooperative body of farmers, has been set up for the purpose. However there are certain lacunae in the functioning of the APMCs. The location of APMC storage facilities is not convenient for many villages. Since they are at a distance, it becomes difficult for small and marginal farmers to take advantage of these. They should open sub-centres within a reasonable distance of say 3 to 5 km from any village so that the facilities can be utilised by all sections of the farmers. The better way would be to periodically collect the produce by the buyer from the village itself. The other improvement needed is; based upon price behaviour in past few years; the initial advance given to the farmer against his produce can he increased from the present level of 60 per cent.

The functioning of some of the APMCs leaves much to be desired. There are instances when farmers are not paid final dues for a long time. This happens either on account of purchasers failing to make payment to APMC or even if APMC receives the payment, they delay payment to the farmers. It becomes difficult for small and marginal farmers to make several trips to APMC for any follow up. The working of such APMCs is therefore to be improved by making their managements accountable for the lapses.

The GOI is planning to amend the APMC Act with a view to enable the farmers to sell their produce in any market. Govt. of India has also circulated to states a model law incorporating a concept of "direct marketing" and contract farming. Some states have already initiated legal or administrative action on this.

5.16. Liberalisation in Agriculture-Establishing Country as a Single Market

The Essential Commodities Act (1955) empowers the Govt. to enforce controls on movement, storage and exports and imports of the essential commodities. These controls were suspended, relaxed and reapplied depending upon the severity of supply shortfalls and abnormal price rise. However there has been some disadvantage of not allowing market to play its role. Further when there are heavy fluctuations in the demand and supply, it takes for the Govt. time to initiate the steps for suspending, relaxing or reapplying the controls and in the meanwhile the situation worsens. In view of this in 2002, Govt. lifted the licensing requirements and movement and storage restrictions on private dealers.

Liberalisation has encompassed most of the sectors and consequently there has been good growth in these sectors. Agriculture too needs to follow suit. India should therefore become single market for all products, especially agricultural produce with free movement of goods throughout the country.

5.17. Diversification in Agricultural Produce-Food Processing-Needed Market Intelligence

The first green revolution had increased the agricultural production to a level which took the country out from a situation of famine and shortage. There is a need to usher in "Second Green Revolution" of value addition in food grain sector. This can be achieved by putting the emphasis on understanding soil characterization, selection of proper fertilizer and water management.

The farmers generally adopt traditional crop pattern. There are varieties of crops which can fetch more money in the market. These include horticulture, floriculture, oilseeds, fruits, vegetables, fish-culture etc. During 2003-04, while India produced 25 million tonnes of edible oil, it imported US$ 2.5 billion tonnes of edible oil. Food/Agro processing if done in rural area will give boost to the rural economy. Presently only 2 per cent of produce is processed. Dairy products also give more income. The crops mentioned above require not only irrigation facility but horticulture and floriculture require polyhouse/greenhouse facility also. Further it is to be noted that processing of food as well as other produce requires capital investment and it can be best done by setting up a cooperative like Amul Dairy at Anand or suitable contract farming.

While there is a need to stress on "market intelligence" based production and consumption system in agriculture, so far there has not been a system for gathering this intelligence and passing on to the farmers with the result farmers continued to produce the crops they were used to. This area needs urgent attention for making agriculture sustainable. The change in the crop pattern will also call for effective extension services as discussed earlier.

There is also a confusion that mars the food sector, with as many as 22 legislations related to it in place at the centre. Obviously there are inter-ministerial differences over the issues. At least three ministries namely health, food and law are directly concerned. Therefore in order to promote the growth of food processing industry, there is an urgent need to frame an integrated food law incorporating all provisions relating to food processing.

5.18. Application of Information Technology (IT) in Agriculture/ Marketing

So long, usage of Information Technology (IT) has been limited to urban area. However recently business software called e-choupal has been developed which offers agriculturists an Internet option by which they can personally participate in a digitally networked rural market. They can even access international market and know the trend of the prices of the farm produce. Since prices in India are likely to follow the international trend, e-choupal can help them to take a decision about the right time when they should sell their produce in the market here. Further they can also learn on-line, the best farm practices, prevailing trends for the crop in the Indian as well as global context, complexities of risk management and the local weather forecast etc.

IT in rural area would bring benefit of knowing the findings of the research in the field of cultivation, which can enable agriculturists to opt for a change from the age-old farming practices. Providing interconnectivity throughout the country is now possible by laying fibre optic cables. IT would thus help rural population not only in applying the developments of the research in their area but also simultaneously contribute in bringing higher form of literacy in them. The same information can also be passed on to the agriculturists through mass media like television, radio and newspapers. India has different climatic zones, temperatures, soil pattern and water availability. In this context by pulling expertise together, the Central Govt. should start television channel exclusively devoted to agriculture which should air programmes related to climate, soil, rain, temperatures, seeds, fertilizers, market and which crops should be sown in different seasons.

Through the project "Village Resource Centre (VRC)- inaugurated on 18th October 2004, on launching of RESOURCESAT, Indian Space Research Organisation (ISRO) plans to set up a number of village resource knowledge centers across the country to provide single window services to rural masses which will include land, revenue, agriculture, weather and loan related information at one place and time.

They would also provide rural communities with dynamic information such as weather and market fluctuations as well as nutrition, health and government entitlements.

5.19. Development of RURBAN Centres

At one end of the paradigm there is over concentration of industries in urban area, at the other end there is no major occupation other than agriculture in the rural area. This has resulted in an unbalanced development in the regions. Whatever we may wish, it is neither possible nor practicable to provide institutional infrastructure necessary for development right in every village. On the other hand the cities are getting so much overpopulated that there is always a serious problem of providing adequate civic amenities. To bridge the gap it is necessary to develop growth centers called RURBAN centers at appropriate places in between the village and the urban area. This will reduce the distance-space and economic-between the village and urban centre" These centers will serve dual purpose as nearest markets for the rural products and ready raw material inputs to small industries. The entrepreneurs can be encouraged to set up at these places small industrial units such as soap making, paper making, electric motor winding, auto repairing, fruit processing, handloom and carpet making etc. There is a need to adopt a deliberate policy by the State to decentralise the industrial growth and develop growth centers nearer to villages.

5.20. Mechanism for Determining Agricultural Costs and Prices

The role of the Commission for Agricultural Costs and Prices (CACP) is to study and recommend to the Govt. of India, Minimum Support Price (MSP) for kharif and rabi crops as well as sugarcane and jute as per a fixed schedule. In line with the recommendations of a high level committee on foodgrains in 2002, Govt. has been announcing the MSP for various crops during the sowing season itself. In 2003-04, production of oilseeds in the country was 25 million tonnes and the country imported US$ 2.5 billion tonnes of edible oil. CACP has therefore recommended higher support prices in 2004-05 for oilseeds in which the country is deficient. Against this background and to stand firmly in the international market, the GOI is contemplating to frame a policy of Crop Plan to be followed by the agriculturists.

CACP thus takes a national overview of various crops while determining the MSP. Unfortunately in practice, many a times the political compulsions as well as bureaucratic intervention have a lot of bearing on the Commission's work.

5.21. Role of Subsidies in Agriculture and When Is the Time Ripe for Removing Them

In India since 65 per cent of agriculture is rain fed and land holdings are largely fragmented, agriculture (especially small and marginal holdings) by itself is not sustainable unless the monsoon is always normal. The natural difficulties of agriculturists are compounded on the one hand by high cost of inputs (like seeds, fertilizers, pesticides, etc.) and on the other hand by lack of good infrastructure (like roads, irrigation facility etc.), cheap and easy credit, storage facility, easy access to markets, remunerative prices for the produce etc. In view of these difficulties, the agriculture sector is heavily subsidised, food subsidy of Rs. 242 billion and fertilizer subsidy of Rs. 110 billion in 2002-03. This phenomenon is not peculiar to India alone but it is so in all other developing as well as developed countries. Only variation could be in the content and the form in which the subsidies are routed.

There are no two opinions about the fact that the public support to agriculture in India is genuinely required and it will not be correct to reduce the present commitment of subsidy unless we route some part of it to more fruitful area. For example, in the matter of giving subsidy to fertilizer, direct benefit to the farmers is only 50-60 per cent and the rest goes to the fertilizer manufacturers. In 2001-02, Govt. announced to rationalize fertilizer pricing and implement the recommendations of the Expenditure Reforms Commission for a phased manner of the price increases, seven per cent per year, and complete decontrol of urea by April 2006.

As regards Minimum Price Support (MSP) mechanism, while it is intended to assure the farmers at least minimum price for their produce, farmers can sometimes face difficulty as well by this mechanism. For example, if the assured price support leads to high level of procurement by the Govt. and afterwards when the stocks pile up to the full brim, Govt. will not be capable of handling physically huge quantum of produce. Eventually the prices will fall in the open market and this will go against the interest of the farmers.

The other way of supporting the agriculture is by way of increasing the investment in development of infrastructure like roads, irrigation facility, cheap and easy credit, storage facility, easy access to markets etc.

For the present, time is not ripe for any substantial reduction of subsidy to agriculture and it is necessary to balance both the alternatives mentioned above when we decide about reducing the subsidy.

5.22. Effective Crop Insurance

The agriculture in India largely depends upon vagaries of nature especially quantum of rainfall and its spread during the agricultural operations. This has direct bearing on uncertainty in agricultural productivity from year to year basis. To ameliorate the fate of the farmers. National Agriculture Insurance Scheme (NAIS) which insures the yield or crop is in operation since *Rabi* 1999-2000. The existing public sector general insurance companies earlier promoted the scheme. However it is now being promoted by the public sector Agriculture Insurance Company (AIC) incorporated in December 2002. To achieve truly the objective, it should sub serve the needs of the farmers and move towards a sustainable acturial regime.

The present scheme has however not become popular with the farmers. The main reason for this is the criteria which has to be fulfilled for the farmers to become eligible for the payment. The scheme provides for payments in the circumstance of natural calamities only and also the insurance cover to the crop is 60 per cent. It does not take into account the market fluctuations with respect to demand as well as prices which are putting the farmers to a great disadvantage. The scheme is more or less cornered by the big farmers and the horticulturists. After harvesting when the new crop comes into the market, most of the small and marginal farmers have to succumb to distress sale as they have to make repayment of the loan as well as to provide for the family expenses.

A new scheme named Farm Income Insurance Scheme has been started as a pilot project. An attempt has been made to incorporate certain modifications to the above scheme. The scheme is under implementation in 19 districts of 12 states for wheat and rice from rabi 2003-04 season. The scheme covers all farmers-lonee farmers on compulsory basis, and non lonee farmers on voluntary basis. The scheme provides comprehensive risk insurance against loss in farm income in a notified area arising out of adverse fluctuations in yield due to natural perils or adverse fluctuation of market price as measured against Minimum Support Price (MSP) or both. The premium rate charged is on actuarial basis. However, there is a subsidy in premium by the Govt. to the extent of 75 per cent to small/marginal farmers and 50 per cent to other farmers. The claim becomes payable if the actual income (AI) for the season falls short of guaranteed income (GI). Actual income is calculated by multiplying current season's actual yield with season's market price. Guaranteed income is average yield of past 7 years multiplied by indemnity level (80% or 90%) multiplied by Minimum Support Price (MSP) of current year. The claims, if any, are automatically calculated based on yield and market price data supplied by State Govt. In this scheme farmers are not required to lodge claim. The claim amount wall be credited to farmer's bank account directly.

The above scheme also suffers from the factor that the guaranteed income is average yield of past 7 years. It is generally seen that there are more number of low yield years than high yield years and hence the farmers are not likely to get much benefit if the average yield of 7 years is considered.

Then there is a proposal of Varsha Bima Yojana, weather insurance scheme. The Agricultural Insurance Company, incorporated in December 2002, is introducing the scheme on a trial basis in 20 rain gauge stations. The pricing policy submitted by the National Insurance Academy (NIA), Pune, which designed the scheme, calls for a topography-specific premium. It seeks to provide cover on five counts: seasonal rainfall cover, sub-optimal distribution of rainfall cover, composite cover, sowing season cover, and catastrophic (drought, mudslides, etc.) cover.

Finally the choice will have to fall upon the scheme, which is best suited to both-the Insurer and the Insured. The scheme should not be such that the farmer has to spend much money and loose many days to recover his claim. In order to bring the insurance culture to the farmers community, the scheme chosen has to be attractive and hassle-free.

5.23. Near Accurate Forecasting Monsoon, Floods and Droughts

Near accurate forecasting of monsoon, floods and drought is central to agriculture which provides livelihood to about 65-70 per cent of the country's total population. It can be helpful to guide farmers on alternate cropping strategies. However the predictions of Indian Meteorological Department (IMD) so far have been of mixed experiences. Despite considerable technological advances in meteorology and satellite communications, the accurate predictions have remained only a distant hope. This appears to be due to the fact that on the one hand the phenomenon is dependent on number of strange factors that range from the El Nino seen off the South American coast to the cyclones of South China Sea and on the other hand is the kind of inadequate equipment which is available at IMD. IMD has been forecasting monsoons based upon its 16-parameter model known as the Parametric Power Regression Model and rain forecasts are made twice in the reason there by giving enough time to the Govt. and the farmers to plan their activities.

As of now IMD is very poorly equipped. The country does not have a single automatic rain-gauge station; eight of 10 cyclone monitoring radars and all 35 of its storm-detection radars are obsolete. It has just 182 surface observatories to collect ground data when there is a need of at least 1,850. It needs 5,000 satellite based automatic rain-gauge stations.

In this regard, China is much ahead in the field of meteorology. China Meteorological Administration (CMA) is in the process of establishing a center for Climate Change with sophisticated modeling facilities.

Therefore, to improve the present situation and for real-time local area forecasts, the Govt. should invest more funds in providing necessary infrastructure, and computing facilities at the IMD.

5.24. Geo-Referencing Project

Geo-Referencing helps in locating a particular village from among all the villages in the State and provide information on the computer about the status of its land. The database is immensely useful for micro-level planning and efficient monitoring of development plans. It can be used for many purposes like watershed development works, to estimate the irrigation command and catchment area, to estimate the produce-wise harvest and even to assess how many farmlands have been flooded or the damage wrecked by the rains or by the hailstorms. Maharashtra has completed this project recently which can be replicated by other states as well. The project involved digitization of all village maps covering 350 taluks in 33 districts supplied by the settlement commissioner and the land records department. It is essentially a superimposition of cadastral land maps on satellite imagery and geo-referencing the digital maps with the help of high resolution satellite data of villages made available through National Remote Sensing Agency of the Department of space. The geo-referencing is then validated with confirmation of actual ground control points. The information can be made available on the State's central server or the Internet so that any department can use it.

The important feature of a project like this one is that the data has to be updated/upgraded with the actual physical status of the land at regular intervals, which normally is not ensured.

5.25. Disaster Management

The tsunami that struck South Asian countries on 26th December 2004 claiming over 1,50,000 human lives is the worst natural disaster to have occurred in contemporary living memory. Besides millions of people becoming homeless, innumerable livestock, wildlife and property perished in the tragedy. India itself has suffered as loss of more than 10,000 human beings, besides loss of livestock, wild animals and property. It is pertinent to note that maximum loss has occurred to fishermen and rural people who were living near the seacoast for their livelihood. Earlier there was large-scale devastation during earthquake in Latur (Maharashtra) and Gujarat.

In India at one end of the paradigm there are conditions of drought due to scarcity of water, at the other end there are floods devastating large areas especially in Bramhaputra basin. Further in coastal area of Orissa and West Bengal there is always severe drainage problem with the result that the area gets inundated by the water which causes heavy loss of crops, cattle and houses. Such disasters are invariably followed by the diseases. The extent of destruction caused during such calamities reflect on the lack of adequate planning and preparedness by the authorities concerned. Although there is no complete remedy against the fury of nature, its effects can be minimised by taking suitable measures. For example, states like Tamil Nadu, Andhra Pradesh, Orissa, and West Bengal which are lashed every monsoon by killer winds and waves from the Bay of Bengal are yet to put cyclone shelters in place.

The measures include pre-emptive measures and post disaster management. In this context there is a need for the country to make suitable legislation to incorporate suitable disaster management measures. A way back in 2001-2002, a Task Force on Disaster Management was constituted at the Centre to take a view on all aspects of the disaster. In this context, GOI has accepted certain recommendations made by the national Committee on Disaster Management including setting up Disaster Management Authority is also on the anvil. Needless to say, there is an urgent need to make suitable legislation laying down policy to tackle the post-disaster situation in a systematic manner. In Railways and Mining Organisations there are detailed well laid down procedures about what is to be done in times of accidents. Likewise there should be a permanent well laid down procedure to meet other disasters.

The advance planning and the arrangements for meeting the disaster, when the event has occurred, can be done at the state or the local level. For managing the situation arisen due to the disaster, decentralisation of the operations should be the key approach. To mitigate the situation, communities should be made timely aware about the risks and they should be involved in designing and implementing the necessary relief operations including resettlement of affected families.

Tsunamis and earthquakes know no national boundaries and it is quite expensive to put in place sensing and monitoring stations at far-flung places by individual countries. It is therefore in the interest of the countries in the region to pull their resources together for disaster control. The SAARC charter already cites science & technology and meteorology as core areas for cooperation. The regional cooperation and sharing of information can further be extended to include SAARC and ASEAN countries, which arrangement can be more effective in tracking the disasters due to proximity of these nations.

5.26. Social Security for Rural Labour

Rural population also consists of a large number of landless people who are below poverty line. There is also undue pressure on land which is not of unlimited measure. Therefore avenues have to be found of such a nature that the rural labour is productively used throughout the year. In this context, rural industrialisation, provision of institutional credit with supporting services for permanent income generating schemes, payment of minimum wages to agricultural labour, providing social security measures for rural poor, etc. need greater attention of the Govt. The National Rural Employment Guarantee Bill 2004 introduced in Parliament on 23.12.2005 is also likely to help rural labour to get the employment at least for 100 days in a financial year.

5.27. Reports of Taking Recourse To Extreme Step (Suicide) by Farmers

Reports do appear in the media about the instances of taking extreme step by the farmers and attributing those to poor crop yield or too much indebtedness. There are many reasons which are attributable to taking extreme step of suicide. These are illness, old age, family problems, persistent stress, extreme poverty and host of specific and non-specific minor as well as major causes. Needless to say, an individual's response towards any problem in life, rather than problem itself, also matters much. Therefore it would not be correct to attribute any instance of taking extreme step to one particular cause without going into the details of each such case. While some cases of taking extreme step due to total crop failure or coercion by the private lenders for returning the loan do occur, any blanket attribution would rather be disparaging to the courage and morale of the average farmer who works in relatively difficult circumstances and who has given his best account during freedom movement of our country.

One can however safely draw a conclusion that the matters relating to rural development cannot be treated lightly and these need to be addressed in all earnestness. If poverty is the cause of such instances, the extremist organisations too derive strength from such conditions.

There are no easy ways to tackle this situation because due to rising costs of inputs and comparatively low prices of the produce, agriculture by itself is tending to become an uneconomic occupation save: the years when there is satisfactory rainfall; those farmers who possess large land holdings; those who are fortunate to have some kind of irrigation facility; those who are having side business like cattle rearing or milk producing etc. Permanent solution for alleviating poverty lies in adopting the strategies prominently mentioned at the end of each chapter in this book. Importantly cheap and easy credit will dissuade the farmers approaching private lenders who charge interest as high as varying from 36 to 120 per cent and use coercive means to recover the loan.

5.28. *Capital Formation in Rural Sector Essential for Overall Economic Development*

It is a wrong notion that the country only spends money on rural development and there is no economic contribution from the rural area. It supplies food to the growing population of the country. It supplies fibre and raw materials to the industries for further processing. Further, it supplies its surplus labour to urban area and industries. Also, industries providing inputs to the agriculture like fertilizers, pesticides, pump sets, tractors, power tillers, threshers etc. thrive due to agriculture only.

It would not be wrong to say that agriculture alone produces real economic surplus over unjust costs of production as against manufacturing and commerce which involve enhancement of raw material inputs only enough to pay for the employees and obtain socially somewhat unjustified profit.

Though faced with many difficulties, India is in a position to export food and other agricultural products to other countries. The agriculture sector contributes 14.7 per cent of the total exports from India. In global scenario presently India is world's second largest producer of wheat and rice, world's third largest exporter of rice and world's seventh largest exporter of wheat.

In less developed country like India, there is a strong relationship between the agricultural growth and the GDP growth. Fig. 3.3 shows the indices of the growth of population, GDP and foodgrains' production from 1950 to 2001.

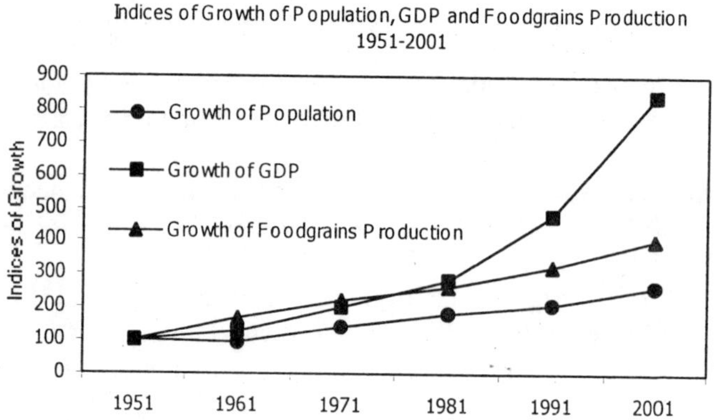

Source: India 2004, (GOI Publication)

Fig. 3.3: Indices of Growth of Population, GDP and Foodgrains Production 1951-2001

It may be seen from the above fig that there is a direct relationship between the growth of foodgrain production and growth of GDP. There is a steady increase in the foodgrain production from 1951 onwards barring the years in which the rainfall was deficient. Needless to highlight, accelerated growth in agriculture sector is very crucial for sustaining the rate of population growth and raising the rate of GDP growth.

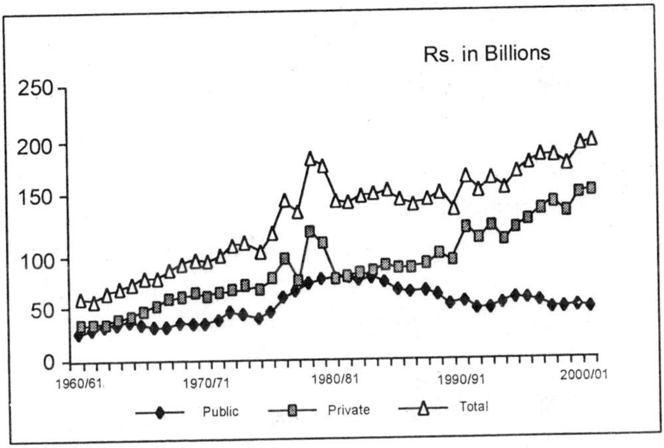

Source: MOA(GOI)/India Sustaining Reform 2003

Fig. 3.4: Gross Capital Formation in Agriculture and Allied Sectors (in Terms of 1993-94 Rupees.)

It is pertinent to note that the public investment in agriculture is gradually declining. The share of public investment in annual agricultural gross capital formation which was 44 per cent in,1985-86 came down to 23 per cent in 2000-01. As against this private investment is gradually increasing. In the long run, less public investment in agriculture sector is likely to affect overall growth rate as in absence of public investment in irrigation, power etc, private investment in farm equipment etc is also likely to be affected.

5.29. Corporatisation of Agriculture

Corporates run the industry. The protagonists of the corporatisation of agriculture argue that it will bring economy of scale in-investment in agriculture, process of production, storage and marketing and adoption of advanced farm technology so as to bring cost effectiveness and competitiveness of Indian produce in the international market. The critics however argue that the move is fraught with certain dangers like changing the present social equilibrium in rural area, depriving entrepreneurial opportunities and initiatives which are presently with individual farmers, reduction in the employment opportunities for the agricultural labour due to employment of mass production techniques etc. Also it is not certain that industry will do agriculture better because in many cases it is found that small farmers give more returns per acre than big ones.

ITC is one of the few corporates who have entered into the field of agriculture. They provide assistance in agricultural operations as well as marketing and the initial response of agriculturists to their entry is positive.

In view of the strong arguments both in favour and against Corporatisation of agriculture, it is advisable to weigh the pros and cons fully before encouraging Corporates to enter the field in a big way. The better way would be if the farmers and the farm labour themselves adopt the work culture of the Corporates or form the efficient operatives and run the agriculture as an industry.

5.30. Taxing Agricultural Income-Pros and Cons

As per Central Income Tax rules, agriculture income as such is exempted from direct income tax. If however a person is having other source of income and also agriculture income, then income from the agriculture as a result of superimposition over income from other source may result in changing the tax slab to higher one. As regards state taxes, these vary widely from state to state. While some of the states collect land revenue, others collect land cess which is quite nominal. In ultimate analysis land revenue/cess collected from the agriculture works out to be very small part of the total revenue receipts, may be about 2 to 3 per cent.

While considering imposition of direct central income tax on agriculture, number of factors will have to be taken into account. There is no doubt that some from elite class, by way of showing income from other sources as income from agriculture, is unscrupulously using a provision of exemption of agriculture from direct income tax. There are also farmers who have large tracts of land or irrigation facility and are earning good income from the agriculture. In both cases mentioned above, from the view point of social justice, levying direct income tax becomes not only justified but a necessary step.

There are however other sides to this issue, (i) Unlike urban income, the farm production and income are beset with many strong uncertainties like weather, crop disease, market demand etc. This makes application of normal principles of taxation to agriculture difficult, (ii) About 80 per cent the farmers fall into the category of small (holding between 1 to 2 ha) and marginal (holding upto 1 ha) farmers and the rest fall in the medium level farmers (owning 2 ha to 10 ha) and higher level farmers (owning more than 10 ha). Small and marginal farmers and majority of the medium level farmers are having marginal or subsistence level income. Thus only those who are having irrigation facility and belong to high level and a few of the medium level may only come in the tax net. (iii) Mostly the agriculturists are not well educated and there is likely to be exploitation of these uneducated class at the hands of some of the unscrupulous staff of the Income Tax Department and this will add one more front for rural people to make compulsory trips to towns, the same way as they are presently required to do for attending to Court or *Kacheri* (iv) The Govt. will have to spend money to support organisation for collection of tax and its administration.

In China, Russia and Japan, income from agriculture is taxable. In this context, it is worthwhile to study the results of the system in China as India is similarly placed in terms of population to be served.

5.31. Maintaining Data Bank

So far most of the data and records of revenue department which include individual land holding, land rent, tax etc were being maintained manually. Information and Technology (IT) has revolutionised the whole system of data keeping. India being one of the leading countries in the world in IT sector, we must focus on maintaining a Data Bank at the block as well as district level. Some of the states have already started the work in this direction. All the land and revenue records, have to be computerised as early as possible so that they are accessible to farmers easily. There is a less chance of manipulation of records in the new system as the parallel record will be maintained at centralised place also. The data to be collected will have to be extended to areas like topography, land holding and ownership, tenancy, land rent, rainfall, crop yield, quantum of irrigation, credit supply, productivity, input cost, demand in market, prices etc. Along with compilation of the data, there should be regular updating of the data, otherwise people will lose faith in the system itself.

6. Adequate Allocation of Financial Resources to Rural/ Agriculture Sector

After independence, efforts have been made to put in place institutional framework for giving impetus to rural development. Initially the State alone played a role in various activities connected with rural development. This was followed by establishment of cooperative sector, later by nationalised banks and now by private sector. There is no denying the fact that the agriculture sector requires massive investments. Such investments have to be through enhanced public investment and credit enabled private investment. The role of each sector is supposed to compliment the role of others. However it is disturbing to note that with the advent of nationalised banks and now private sector, cooperative sector is getting marginalised. It will not be correct to find fault with the concept of cooperation. In fact since its inception it has played a key role in the past in rural development. However this sector has suffered on account of bureaucratic control and leadership of cooperative institutions developing vested intérest over a period of time. On the other hand, Amul Dairy at Anand in Gujarat is one of the finest examples of successful cooperative sector even now, which has been recognised world over as one of the best performers. This example is worth emulating by others.

There is a debate going on in the country about the nature and quantum of subsidy, a State should provide to agriculture and setting up industries in rural sector. So long as the farmers do not get a return from the agriculture for at least minimum sustenance level for meeting the expenditure on food, education, healthcare and housing, the State will have to provide interventions in the form of subsidies to rural sector. All the countries whether developed or under developed are providing subsidies to this sector. The degree and form of support may vary. In this respect "Non-Trade Support" will be a more suitable nomenclature than calling it as a "Subsidy to Agriculture".

Fig 3.5 shows financial outlays on the schemes of the Ministry of Rural Development 2003-04.

Rs. Crore

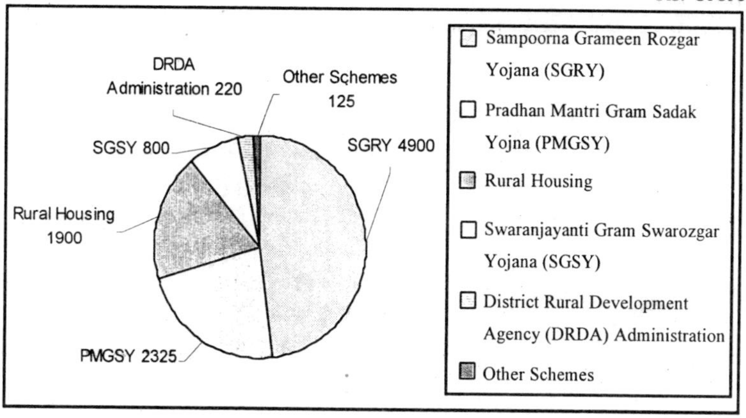

Source: NIRD Rural Development Statistics, 2002-03

Fig. 3.5: Financial Outlay on the schemes of the Minstry of Rural Development 2002-03

It will be seen that the outlay of Rs. 9,925 crore on 4 schemes namely Sampoorna Grameen Rozgar Yojana, Pradhan Mantri Gram Sadak Yojana, Rural Housing and Swarnajayanti Gram Swarozgar Yojana alone constitutes about 97 per cent of the total outlay of Rs. 10,270 crore. The allocation to Rural Development Schemes in 2005-06 (BE) is Rs. 18,334 crore. The outlay in future will have to be enhanced substantially in the context of National Rural Employment Guarantee Bill 2004 which has been introduced in the Parliament on 23rd December 2004

Agriculture being a state subject, the bulk of the public investment in agriculture takes place at the state level, and the Central Government's support to states acts as a catalyst.

Table 3.15 gives planwise total public outlay, total expenditure on agriculture and percentage of agriculture expenditure in total public outlay. It will be seen from the table that even though there has been increase in allocation in absolute term to agriculture & allied activities and irrigation, there is a steady decline in the allocation to agriculture as a percentage of the total public outlay from 32 per cent in the First Plan (1951-56) to 10.6 per cent in Tenth Plan (1997-2002).

In recent years (2001-02 and 2002-03), the investment per year on agriculture and allied activities and irrigation amounted to about 2 per

Table 3.15: Agriculture and Five Year Plans (1951-56 to 2002-07)

Rs. Crore

Outlay/Expenditure	*First Plan 1951-56*	*Second Plan 1956-61*	*Third Plan 1961-66*	*Annual Plan 1966-69*	*Fourth Plan 1969-74*	*Fifth Plan 1974-*	*Sixth Plan 1980-85*	*Seventh Plan 1985-90*	*Eighth Plan 1992-97*	*Ninth Plan 1997-2002*	*Tenth Plan 1997-2002*
Total Public Outlay	2,378	4,800	8,577	6,665	15,902	39,303	97,500	1,80,000	4,34,100	8,59,169	15,25,639
Agriculture and Allied Activities	360	565	1,089	1,051	2,385	—	5,695	10,524	22,467	42,462	58,933
Irrigation and Food Control	400	458	664	445	1,087		12,160	16,979	32,525	55,420	1,03,315
Total Expenditure on Agriculture	760	1,023	1,753	1,496	3,472	8,253	17,855	27,503	54,992	97.882	1,62,248
Percentage of Agriculture expenditure in Total Public Outlay	32	21	20	22	22	21	18	15	13	11.4	10.6

Source: Five Year Plan Reports/Statistical Abstract-India, 2003(GOI)

cent of GDP, whereas agriculture and allied products contributed to 21 per cent of GDP (2003-04) and 14.7 per cent of the country's total export earnings.

The growth rate in agricultural GDP vis-à-vis Total GDP during recent years is given in Table 3.16.

Table 3.16: Growth Rate in Total GDP and Agriculture and Allied Sector from 2000-01 to 2004-05

Sl No.	Growth Rate in	% growth in GDP over the Previous Year				
		2000-01	*2001-02*	*2002-03*	*2003-04*	*2004-05*
1.	Total GDP	4.4	5.8	4.0	8.5	6.9
2.	Agriculture and allied sector (in terms of GDP)	(–)0.1	6.3	(–)5.2	9.6	1.1

Source: Economic Survey (GOI) from 2000-2001 to 2004-05.

Less investment is one of the major handicaps for not achieving good growth rate in agriculture as well as the originally set objectives in any field, may it be, rural development, education, healthcare, poverty alleviation, etc. The other major reason is that the end person for whose benefit all the schemes were designed did not actually get any significant benefit. Former Prime Minister Mr. Rajiv Gandhi rightly observed that out of every Rs 100/- spent by the Govt., only Rs 15/- reaches the end person. The rest is being spent on maintaining the administering departments and drained away by the vested interests. The remedy lies in (1) increasing the investment in rural sector to 20 per cent of the total public outlay in place of 10.6 per cent, (2) decentralisation of the planning process and involving rural people in the planning and (3) giving full autonomy to rural bodies for implementing the schemes.

As far as two main components, namely basic education and primary healthcare, of rural development are concerned, it has been suggested that if additional financial resources are required to bridge the gap between financial resources required and resources available, they should be found from inter-sectoral adjustments by transferring resources from less priority sectors. If there still remains a gap, cess on cental taxes may be levied to an extent to fill up the gap subject to certain provisions mentioned in the chapter on Budget in this book. In fact such a cess or surcharge has been levied in the past to meet certain exigencies of the situations.

7. Managing External Factors-World Trade Organisation (WTO)

7.1 Main Features of WTO

7.1.1. Objective

The main objective of WTO, earlier known by General Agreement on Trade and Tariff (GATT), is to facilitate international trade by restricting the tariff and other barriers so as to enable liberal flow of goods and services throughout the world. GATT which is also known as Dunkel proposal was signed in 1994. Arthur Dunkel was at that time Director General of GATT and played very important role in bringing about consensus between the countries which culminated in signing the agreement by 124 countries including India. WTO succeeded GATT with effect from 1st January, 1995.

7.1.2. Scope

(i) WTO deals with trade in manufactured goods as well as agricultural products.

(ii) Trade Related Intellectual Property Rights (TRIPS): The intellectual property under TRIPS covers copy rights, trade marks, trade secrets, geographical indication, industrial designs, layout designs of integrated circuits, patents, micro organisms and plant varieties with sui generis system i.e. a unique system of its own. Earlier only process was patented. Now product is also included. The royalty will have to be paid for export of the patented product.

(iii) Trade Related Investment Measures (TRIMS): The objective of TRIMS is to ensure free flow of capital from one country to another without conditionality of use of local materials or exports obligation under WTO.

(iv) Services: like telecommunications, banking, insurance etc. have since been included.

7.2 Impact of WTO on Indian Agriculture

WTO deals with trade in manufactured goods as well as agricultural products. It will have both favourable as well as unfavourable effect on Indian agriculture in the following manner.

7.2.1. Subsidy

There are three types of subsidies namely (i) input subsidy, (ii) export subsidy and (iii) Govt programmes and services. Aggregate Measurement of Support (AMS) is the basis of arriving at the subsidies. AMS comprises both product and non-product specific support. Product specific support is calculated as the difference between the domestic price of the related agricultural product and the fixed reference price multiplied by the quantity of output eligible to receive the domestic support. The non-product specific support is calculated in terms of subsidy given to agriculture for inputs of fertilizer, canal irrigation, electricity, seed and credit.

As per GATT, there is a cap on subsidies to agriculture to the extent of 10 per cent of the total value of the output. So far India has not exceeded this limit. However if increases in procurement prices for the produce and subsidy in inputs to agriculture continue, after a period of time the limit of 10 per cent may be crossed. On the other hand if developed countries do not agree to reduce the present level of subsidy given to their farmers, it would be difficult for underdeveloped country like India to export the surplus produce to developed countries like USA or Europe.

7.2.2. Seed Patenting

As per TRIPS, agriculture related microorganisms and microbiological processes are patentable, whereas as per Indian law these are not patentable. However farmer can use seed produced by him in subsequent crops, but he is not supposed to sell it as a commercial venture. This activity can be undertaken by the plant breeders who will be protected under Plant Breeders Right (PBR). Researcher can use one protected variety to breed another new variety. Under this regime seeds produced by Indian Agriculture Research Institute will not have any difficulty. Bulk of the seed requirement in India in an/case is met from the seed produced by the Indian farmers.

7.2.3. Import of Agricultural Products

As per WTO agreement, a country will have to import agricultural products to the extent of 3 per cent of its requirement. Presently India does not have to follow this as it gets benefit due to another enabling provision that is related to balance of payment (BOP). So long as BOP position is negative, the above condition is not applicable to India. However if our aim is to export agricultural products we will have to plan our strategy in a manner: (i) produce the product for which there is a demand in the international market, (ii) improve the quality to the level of international standard and (iii) production has to be so cost effective that its price is competitive in the international market. Such a situation can be obtained when farmers adopt new farm technology extensively and there is adequate guidance and support from the Govt. to the agriculture sector in the form of effective Extension Services.

7.2.4. Public Distribution System

The food distributed through Public Distribution System involves subsidy and the same comes in the perview of WTO. In India this subsidy on food is well justified as 26 per cent of the country's total population is still below poverty line (BPL). However, exclusion of higher income groups from PDS can be considered which will reduce the subsidy burden on the Govt. as well. Although PDS is very well intentioned, the system is being misused by unscrupulous elements such that in some places it has been observed that the food from PDS is finding its way in the open market where it is sold at higher prices. These loopholes have to be plugged by way of proper vigilance over implementation of the scheme.

7.3 Proceedings of Last World Trade Organisation Conference held on 10th to 14th September, 2003

Conference of WTO was held at Cancun in Mexico from 10th to 14th September 2003 in which 146 member countries participated. This conference was preceded by the last conference in Doha which dealt with agriculture and other issues. Some issues which pertained to investment and discussed at Singapore were also to be taken up. These included transparency, competition, trade facilitation and frame work for Govt. procurement. India obviously was more concerned about taking up the issue of agriculture especially about removal of heavy subsidy presently being given by developed countries especially USA and EU countries to their farmers which has made export of agricultural produce from developing country like India difficult. On the other hand the developed countries did not want to agree for reduction of subsidies being given to their farmers since they thought such a move would lead to import of produce from developing countries thus affecting their farmers. Vain attempt was made to simultaneously raise the investment issues of Singapore, but G-21 led by India, Brazil and China insisted on first resolving the issue of agriculture. Ultimately the conference ended in a deadlock over the farm subsidy issue, which meant in plain terms the conference ended in a failure. In any forum, unless countries adopt to give and take policy, such conferences are bound to be unsuccessful. Further in democratically elected countries, it is difficult for the Governments to take a position which may antagonize large section of their population.

There is also a benefit in going for regional grouping on the lines of EU. We already have SAARC but unfortunately differences which exist on political front between the countries are not allowing growth of regional cooperation. There is also another regional grouping in our neighbourhood namely ASEAN.

7.4 WTO Trade Talks at Geneva-July 2004

The main sticking point in past trade negotiations among the 147 member nations is farm trade. The rich countries subsidise farmers, as do most poorer nations. However, richer nations insist that the developing nations should open up markets and drop sops faster than they (rich nations) do. During a decade long marathon Uruguay Round trade negotiations, rich nations made certain commitments to phase out subsidies and opening up market. Instead of keeping their commitments, every time they come for negotiations, they offer what they were supposed to do years ago in return for more concessions from the developing countries. Against this backdrop, the trade talks that ended on July 31, 2004 in Geneva were also not conclusive. However they have achieved two things, one, they have kept the door open for further negotiations scheduled for 2005 and two, they have made rich nations to concede to important demands of poorer countries like India. Key WTO members, including USA, EU, and Japan, agreed to the elimination of export subsidies, a key developing country demand, at a date yet to be set, and to limit other subsidies, and to lower tariff barriers.

13. IMPACT OF GLOBALISATION ON RURAL ECONOMY

One may like it or not, Globalisation is unavoidable. In today's world no country can afford to remain isolated. Globalisation has brought in its wake awareness in India about the necessity of improvements in productivity, quality, cost effectiveness, and work culture. It has raised income levels even in agricultural community. These are strong positive contributions of globalisation. However globalisation has also brought skewness in the income structure resulting in high imbalance in different sectors. In 1970's, 74 per cent of workforce was engaged in agriculture and contributed 48 per cent of GDP (74/48). The share of manufacturing and services at that time were 11/20 and 15/32. As against this in 2000, picture has dramatically changed to 60/28 in agriculture, 16/25 in manufacturing and 24/47 in services. This indicates that the workforce engaged in manufacturing and services, which are much smaller than agriculture, are associated with huge amount of product value and in turn reaping high benefits. This does not augur well from the view point of overall perspective as it has increased income disparities to a high level. In this context the role of the Govt. for implementing the schemes that promote social justice and equal opportunities assumes great importance.

The views expressed in Discovery of India (1946) (p. 556) about economic model amply reflect Pt. Jawahar Lal Nehru's philosophy where he states: "It is a weakness of modern civilisation that it is progressively going farther away from the life-giving elements. The competitive and acquisitive characteristics of modern capitalistic society, the enthronement of wealth above everything else, the continuous strain and lack of security for many, add to the ill-health of the mind and produce neurotic states. A saner and more balanced economic structure would lead to an improvement of these conditions. Even so it will be necessary to have greater and more living contacts with the land and the nature. This does not mean a return to the land in the old and limited sense of the word, or to a going back to primitive ways of life. That remedy might well be worse than the disease".

Having said earlier that globalisation is inevitable, Indian farmer will be required to maintain quality of produce as well as competitive cost to stand in the international market. The backward linkages like improvement in farming practices and various inputs like availability of quality seeds, fertilizers, irrigation facilities etc. will have to be worked out. The Govt. too will have to provide necessary infrastructure like lab to land, facilities for storage, packing, transportation etc.

14. CONCLUSION

The basic ideas of Gandhiji about rural development are relevant even today. He emphasised on making village, as far as possible, a self-dependent unit and urban people should use, as far as possible, only those products which are produced by the rural population so that the rural economy becomes sustainable. This approach however did not preclude mass production of items that are required for building developmental projects. There is no doubt that there has been progress in the development of rural area after independence but it is far from satisfactory. Villages are still lacking basic living facilities like education to all eligible children, healthcare, housing, safe drinking water, sanitation etc They also lack proper infrastructure like all weather roads, communication facilities etc.

Farmers are still largely dependent on vagaries of nature as there is not much progress in the enhancement of irrigation facilities. Adoption of farm technology can improve the yield from the agriculture. This can however be achieved only through availability of cheap and easy credit facility together with effective extension services to be manned by the sincere and committed extension staff. Since employment in agriculture is seasonal, there is a problem of employment for the youth. Vocational courses should be made a part of the school education curriculum so that youth can take up self-income generation schemes.

The status of rural female population is to be enhanced by way of special efforts in enrolling girls for education in schools, preventing discrimination against women in matters of employment and bring them on par with men in socio-economic political system. There is a need to spread awareness about advantages of having small family and adopting budgetary concept for income and expenditure for the year as a whole to avoid debt burden. Institutional framework has been established to serve the rural population as a whole. However it is observed that while middle and rich farmers have taken advantage of various schemes, not much benefit has reached the tribal, scheduled casts and economically weaker sections of society.

It is observed that the functioning of Panchayat Raj Institutions has been affected due to non-availability of adequate fund and its proper utilisation. Decentralisation of power to rural bodies is essential to make a concept of self-governance by the rural population successful. In this context national consensus will have to be built up for the proposal that the money sanctioned for Panchayats will hereafter go to them from the central Govt. directly, but via the State Govt. with the instruction that the money earmarked for Panchayats has to be used exclusively for the development purpose. It is to be noted that only a small part of the money allocated reaches the person for whom the scheme is actually formulated. It is necessary that only such staff should be engaged in rural area, who are committed to social work and are sensitive to the problems being faced especially by the rural poor.

It is a matter of great concern that the investment in agriculture as a percentage of total plan outlay is steadily decreasing in spite of the fact that agriculture by itself is not self-sustaining and not even basic living facilities are created in large part of rural area.

Since development is directly related to resources, this trend has to be corrected. Some people advocate either reducing or removing subsidy altogether being given to inputs to agriculture and allied sector and village and small industries. They also profess imposition of direct central tax on agricultural income. These issues have to be viewed in the larger context of prevalence of abject poverty in large section of the rural population and national commitment to social justice and equal opportunities to all citizens.

4

Primary Healthcare in Rural Area: A Perspective

1. INTRODUCTION TO PRIMARY HEALTHCARE

1. Priority of Healthcare

There is a proverb in English, "Health is Wealth". In Indian scriptures also the importance of health has been highlighted. Good health is an essential component for individual happiness and progress and collectively human productivity and development. In view of this health services have been given an important place in the fundamental rights. India is however far away from realising the goal of providing basic healthcare to all. To quote Nobel laureate Dr Amartya Sen, illiteracy, malnutrition and lack of adequate healthcare are three deadly enemies of a just society. Presently our country's population is 1028.7 million which accounts for one sixth of the world's total population. 26 per cent of our country's population is still below poverty line and large majority of BPL families live in rural area. The living conditions of these people are miserable. Poverty is direct cause for diseases and ailments. These people do not have even means to meet basic necessities like food, education, healthcare, housing etc. India is producing food more than its need. However there are lacunae in public distribution system (PDS) because of which food is not reaching to needy. If PDS is streamlined, there will be no case of starvation anywhere. There are many priorities before the State but providing basic education and healthcare to everybody occupies topmost priorities. The idea is, if basic education and healthcare are taken care of, the rest of the development would automatically follow. After independence the State had to involve itself in many areas including infrastructure development as there were no private entrepreneurs, barring a few, to take up these projects. It is now a considered opinion that the State should: (i) gradually withdraw from the activities of business areas, (ii) fully concentrate on providing education and healthcare to all and (iii) use State's scarce resources first for meeting the goals of education and healthcare. In this Chapter 4 of the book we would be dealing with healthcare which is top most priority along with education. Although healthcare should comprise both preventive aspect as well as treatment of diseases, obviously emphasis has to be laid preferably on preventive care.

2. Definition of Health

World Health Organisation (WHO) in 1948 has defined health as follows:

"Health is a state of complete physical, mental and social wellbeing and not merely an absence of disease or infirmity." The definition has since been enlarged to include "ability to lead a socially and economically productive life."

3. Human Development Index (HDI)

As already mentioned in Part-2 on Basic Education that the Human Development Index as a valid measure of country's progress is calculated as an average of the sum of three indices namely longevity, educational attainment and income. Here longevity is represented by the factor-life expectancy at birth. Since life expectancy at birth is directly related to public healthcare available in the country, great importance is attached to the latter. The access and the quality of healthcare are more relevant to rural area than urban area as in rural area there is no option like going to private professional doctors first because such professionals do not reside in villages and secondly the charges of these doctors and medicines prescribed by them are beyond the reach of rural population.

2. CONSTITUTIONAL POSITION & NATIONAL HEALTH POLICIES

1. Right to Health

Health issues and Human rights are inseparable. The directive principles of the Indian Constitution have been interpreted by the highest judiciary in India as providing right to health and right to food as rights enforceable through a court of law. As per Constitution of India, public health and sanitation comes under State list and population control and family planning comes under concurrent list. Further it has been provided under articles 41, 42 and 47 of the constitution that:

"The State shall, within the limits of its economic capacity and development, make effective provision for securing the right to work, to education and to public assistance in case of unemployment, old age, sickness and disablement, and in other case of underserved want.

The State shall endeavour to raise the level of nutrition and standard of living to improve public health.

Right to humane conditions of work and maternity relief."

The constitutional guarantees and fundamental rights must reach the most needy sections and unfortunate incidents like starvation deaths, malnutrition and epidemic deaths should be prevented by putting in place effective health care system and concerted efforts of the public spirited individuals and community itself.

The right to health also finds place in the preamble of WHO constitution where it is mentioned that it is one of the fundamental rights of every human being to enjoy the highest attainable standard of health.

India is a signatory to Alma-Ata Declaration adopted by WHO-UNICEF International Conference held in 1978, which was attended by 134 countries. The conference called for working towards WHO goal of Health for All by 2000 and declared primary health care as a way to achieving it.

WHO defined Health for All as, "attainment of a level of health that will enable every individual to lead a socially and economically productive life."

2. National Health Policies-1983 and 2002

The National Health Policy-1983 affirmed commitment to the goal of Health for All by 2002. Major targets to be achieved by the year 2000 as set in the policy were: Reduction of infant mortality rate from 125 (1978) to below 60; Reduction of crude birth rate from 33/1000 to 21; Reduction of crude death rate from 14/1000 to 9; Providing drinking water to all rural population.

As against the above targets, the actual achievements as in the year 2002 have been: Infant mortality rate 67; Crude birth rate 25 and Crude death rate 8.1. Although some progress has been made in providing drinking water in rural area, there are still number of villages where this basic facility is not available either in the village itself or within a reasonable distance.

National Health Policy-2002 has now laid down specific goals to be achieved by the year 2015 as follows:

Eradicate Polio	2005
Eliminate Leprosy	2005
Eliminate Kala Azar	2010
Eliminate Lymphatic Filariasis	2015
Achieve zero level growth of HIV/AIDS	2007
Reduce mortality by 50 per cent on account of TB, Malaria and other air and water born diseases	2010
Reduce prevalence of blindness to 0.5 per cent	2010
Reduce Infant Mortality Rate to 30/1000 and Maternal Mortality Rate to 100/Lakh	2010
Increase utilisation of Public Health facilities from current level of <20 per cent to >75 per cent	2010
Establish an integrated system of Surveillance, National Health Accounts and Health Statistics	2005
Increase health expenditure by Govt as a percentage of GDP from the existing 0.9 per cent to 2 per cent	2010
Increase share of central grants to constitute at least 25 per cent of total health spending	2010
Increase state sector health spending from 5.5 per cent to 7 per cent of the budget	2005
Further increase to 8 per cent of the budget	2010

3. REPORTS OF COMMITTEES ON PRIMARY HEALTHCARE

In past various Committees were formed from time to time to review the then existing health system and recommend measures for improvement. In this context the gist of the recommendations of the earliest Committees namely Bhore Committee and Shrivastav Committee are given here.

1. Bhore Committee (1946)

The Committee named Health Survey and Development Committee with Mr. Joseph Bhore as Chairman submitted first comprehensive report in 1946 on the health situation existing then and the recommendations for development of the health services. The Committee recommended that the health programme should lay emphasis on the preventive aspect of the health and these activities should go alongside those concerned with the treatment. The Committee also recommended integration of the preventive and curative services and establishment of Primary Health Centres in the rural area, each PHC as per long term programme covering a population of 10,000 to 20,000. The PHC will have two medical officers beside other paramedical staff.

2. Shrivastav Committee (1975)

A Committee on Medical Education and Support Manpower headed by Mr. Shrivastav submitted its report in 1975. The major recommendation of report is that the health workers from within the community should be trained so as to enable them to provide simple primary i.e. preventive and curative healthcare to that community. This means involvement of people in taking care of their own health. The Committee recommended for providing one male and one female health worker for a population of 5000. The recommendations finally led to the formation of the Rural Health Scheme.

4. COMMUNITY HEALTHCARE

1. Scope

Health is influenced by wide ranging factors such as food, housing, living style, cleanliness of drinking water, hygiene, sanitation etc. It would not suffice to mean it. simply as medical care. In order to achieve desired results in health care of the whole population, multidimensional approach is therefore called for.

2. Concept of Community Health

It is difficult to provide for health promotion exclusively for an individual in the rural area. The service has to be directed towards fulfilling the needs of the community as a whole in which individual gets automatically covered. There are two ways to achieve the objective of community health. One is by way of providing environment in the village which is congenial to good health such as sanitation, safe drinking water, food, nutrition etc. and the second is by providing personal care at the health centers. This has given birth to the idea of providing basic health services through the medium of primary health centres and sub-centres. Likewise community medicine has appeared on the scene. It again shares a common ground – prevention of disease and promotion of health

3. Healthcare Service Levels

The healthcare services are covered at three levels.

3.1 Primary Level

The basic or essential health care is provided at this level. These services are provided through Primary Health Centres and Sub-centres set up in the rural area. This is the first level at which the community gets in contact with the health system.

3.2 Secondary Level

District hospitals and rural health centres form the secondary level at which better medical facilities are available. These also serve as the first referral hospitals.

3.3 Tertiary Level

Medical college hospitals and specialized hospitals come under this level. At this level, highly sophisticated diagnostic equipments as well as more experienced specialists are available. These hospitals are expected to deal with more complex medical cases and also conduct research work.

4. Alma-Ata Declaration on Primary Health Care

Alma-Ata Declaration (1978) highlighted the essential elements of the primary health care as : education regarding prevailing health problems and the methods of preventing and controlling them; providing food and proper nutrition; providing safe drinking water and basic sanitation; maternal and child health care and family planning; immunization against major infectious diseases; prevention and control of local endemic diseases; appropriate treatment of common diseases and injuries; and provision of essential medicines.

It would be seen from the content of the primary health care that it would focus on promoting health services in the fields which are of general concern to the public.

5. STATUS OF THE HEALTH FOR ALL PROGRAMME IN INDIA WITH REFERENCE TO WHO INDICATORS FOR MONITORING THE PROGRAMME

The present status of the Health for All Programme in India with reference to World Health Organisation (1981) Indicators for monitoring the progress of this Programme is discussed here.

1. Health Policy

1.1. Political Commitment to Health for All

In line with various deliberations and guidelines of WHO and Declaration adopted at Alma-Ata International Conference, India formulated National Health Policy in 1983 and subsequently revised in 2002 with the emphasis on providing Health for All. Although cue has been taken by India from the above organisations, the health indicators and their targets for monitoring the progress have been specified taking into account the circumstances as existing in the country. This is all right as far as policy formulation is concerned. As regards implementation of the policy to cover entire population, important factors namely necessary resources, political will and commitment of bureaucracy to achieve the goal of Health for All are found to be inadequate to meet the requirement of the goals. Hence there are serious shortfalls in achieving the targets.

1.2. Resource Allocation

In a developing country like India with population of 1028.7 million, resource is a definite constraint. There are several fields such as health, education, agriculture, transport, power, communication, defence, oil, fertilizers etc. which require funds for development. Since there is a limitation for raising the resources in a developing country as ours every field is facing the resource crunch. Continued losses being suffered by some of the public sector enterprises have only aggravated the problem. Govt. has therefore no alternative but to disinvest/sell such loss making enterprises to private entrepreneurs. Soon Govt. has to dispense with all business activities except pertaining to strategic areas and fully concentrate on providing basic requirements such as education, healthcare, shelter, food safe drinking water etc. In recognition of the high importance of the healthcare, as stated earlier, National Health Policy-2000 stipulates time bound specific fund allocations to this sector such as:

Increase health expenditure by Govt as a percentage of GDP from the existing 0.9 per cent to 2 per cent	2010
Increase share of central grants to constitute at least 25 per cent of total health spending	2010
Increase state sector health spending from 5.5 per cent to 7 per cent of the budget	2005
Further increase to 8 per cent of the budget	2010

In the face of strong pressures from different sectors for allocation of more funds, it is to be ensured that the above targets which pertain to very important field like healthcare are not stifled.

The funds are provided to Health Services from different channels namely Central Govt., State Govt. and Zilla Parishad The entire expenditure on Family Welfare Programme and combating certain diseases are borne by the Central Govt. The payment of salaries of State appointed doctors and staff are provided by the State Govt. and Zilla Parishad appointed staff by the Zilla Parishad. The State Govt.

provides for the facilities. Apart from this, there is aid from international bodies like World Bank, Asian Development Bank etc for specific projects and programmes related to healthcare.

However it is disappointing to note that more than 80 per cent of the financial allocation to Health Services is spent only on payment of salaries of doctors and staff and less than 20 per cent is available for the expenditure on development of facilities. This is one of the main reason for inadequate facilities available in Primary Health Centres and Sub-centres and their low performance.

1.3. Equity of Distribution of Health Services

The institutional health services have come up in urban area only. In particular private doctors and hospitals are generally located in only urban area. The reason is that the paying capacity in urban area is high and the living conditions are also comfortable there. However the services of the private doctors and hospitals, being costly, can be utilised only by those who are having good earning capacity. The rest of the urban population have to depend upon Govt/Municipal hospitals. The problem is more acute in the rural area where the access to health services is most difficult. Since rural population constitutes 72.22 per cent of the total population of India, the issue of equity of distribution of health services assumes great importance. The only way to do this is by decentralising the health services and locating them in the rural area. In doing so, priority will have to be given to preventive and first line curative aspect of the healthcare. This would mean strengthening of the existing Primary Health Centres and Sub-Centres as well as opening new centres in the places where access to services is difficult.

Since major disease burden comprises diseases like malaria, tuberculosis, blindness, leprosy, HIV/AIDS etc., tackling these diseases as well as reduction of infant/maternal mortality rates calls for higher priority.

1.4. Community Involvement

There is a wide spread expectation among the public that the health services should be easily available to the general public and not limited to a few. In the above context, the health services obviously should cover both preventive and curative aspects. There is no way to achieve this gigantic objective by any means other than focusing on developing proper primary health care which is fairly accessible to the community and brining forth involvement of the community itself for promoting their own health. This two pronged approach is more likely to succeed rather than providing dispensaries and hospitals for which there will always be limitations in terms of both financial and physical resources.

Awareness of the elements of health and how to maintain general health are two key factors which have to be explained to the people in the rural area. The function of *health guides* in this context assumes great importance. So also the services of local *Dai* have become important in the village community as many of the delivery related urgent problems are tackled by *Dai* herself.

1.5. Organisational Framework

In our country at the Centre, the nodal Ministry is the Ministry of Health and Family Welfare, Director General of Health Services (DGHS) and Central Council of Health and Family Welfare (CCHFW). DGHS looks after planning and coordination of all health matters in the country. Union Minister is the Chairman and the State Health Ministers are the members of the CCHFW. It acts as a catalyst for consultations between Centre and State and recommends broad outlines of policy in matters of health and family welfare. Indian Medical Council (IMC) ensures that medical colleges and hospitals satisfy certain standards to be eligible to impart medical education. Indian Council for Medical Research (ICMR), apart from conducting research in medical field also frames National Guidelines for Accreditation, Supervision and Regulation of medical practices and clinics in the country. Accreditation is to ensure minimum standard and ethical medical practices.

At the State level, there is a Ministry of Health and Family Welfare and Director of Health and Family Welfare who is principal advisor to the State Govt on all the matters relating to health and family welfare. In some of the states, for looking after medical colleges, there is a Director of Medical Education.

At the district level, there is Chief Medical Officer/District Health Officer as in charge of the district for all the health and family welfare programmes. Primary Health Centre is headed by the Medical Officer and sub-centre, which is rural outpost of the health system, is manned by one male and one female health worker. The last outposts at the village level are Village Health Guide and local trained Dai.

The organisational structure as mentioned above is well defined. However, in implementation it suffers from the typical bureaucratic approach both from the Ministry as well as medical fraternity. There is already paucity of funds for all the sectors including health services. The bureaucratic approach aggravates the problem further as whatever funds are available, these are not being provided to Primary Health Centres and Sub-Centres in time. The centres are not properly equipped and even medicines for common diseases are often not available at the centres. Further, there is laxity of posting even the sanctioned medical staff in the centres, especially doctors whose turnover is high because they always prefer postings in cities.

2. Social and Economic Factors related to Health

2.1. Population Increase

The world's population is about six billion right now and it is growing at quite a faster pace. A study published by the Royal Society, UK concluded that there are 1,000 times too many humans on earth to be sustainable for very much longer considering carbon dioxide production, energy use, biomass consumption and geographical range. In the case of India, the problem is more acute.

As per last census (2001), India's total population is 1028.7 million which constitutes 16.87 percent of the world's population. Rural population constitutes 72.22 percent of the total population of India. There are both positive as well as negative aspects of the growing population. More population means more manpower for the industries and the labour component of the total cost of the product is less. Indian products would thus be more competitive in prices in the global market. Also presuming other factors are same everywhere, the developed countries would prefer to set up their industries and make more investments in India thus resulting in better economic growth of the country. The darker side of the growing population is that whatever development effort has been made so far, could not lead to any significant improvements even in vital areas like basic education, primary healthcare and raising living standard in rural area. On the other hand overpopulation is putting tremendous pressure on the civic amenities in both urban and rural areas. It is even becoming difficult to provide safe drinking water and proper sanitation to the people in general and rural people in particular. There are unwieldy population pressures on the public services like water, power, transport etc. Then there is an acute problem of unemployment. The increasing trend of crimes is also being attributed to increasing rate of unemployment. Thus beyond a point, overpopulation is bound to impact unfavourably on raising the standard of living of a common man.

Table 4.1 shows the relationship of total fertility rate (TFR) with certain background characteristics like literacy, place of residence i.e. urban/rural, religion, caste and standard of living as per National Family Health Survey conducted in India during 1998-1999.

Table 4.1: Total Fertility Rate by Background Characteristics 1998-99

Background Characteristics	*Total Fertility Rate*
Residence	
Urban	2.27
Rural	3.07
Education	
Illiterate	3.47
Literate, < middle school complete	2.64
Middle school complete	2.26
High school complete and above	1.99
Religion	
Hindu	2.78
Muslim	3.59
Christian	2.44
Sikh	2.26
Jain	1.90
Buddhist/Neo-Buddhist	2.13
Other	2.33
No religion	3.91
Caste/tribe	
Scheduled caste	3.15
Scheduled tribe	3.06
Other backward class	2.83
Other	2.66
Standard of living index	
Low	2.37
Medium	2.85
High	2.10
Total	2.85

Source: Park-2005

Note: Total fertility rate (TFR) means average number of children a women would have and the measure of TFR, gives magnitude of 'completed family size'.

While there is fairly good response to family planning in urban area, it is lukewarm in rural area. Since the literal meaning of *family planning* connotes coercion by the State, which in any case in democracy cannot dictate family planning for the people, the term family planning programme has been changed to family welfare programme. In this context the salient features of National Population Policy 2000 are given below.

National Population Policy (NPP)-2000: The NPP-2000 has identified the immediate objectives as meeting the unmet needs for contraception, healthcare infrastructure and trained health personnel and to provide integrated service delivery, with the objectives: (i) Strengthen community health centres, primary health centres and sub-centres; (ii) Augment skills of health personnel and healthcare providers; and (iii) Bring about convergence in the implementation of related social sector programme so that the Family Welfare programme becomes a people's programme; The NPP enunciates certain socio-demographic goals to be achieved by 2010, which are expected to lead to achieving population stabilization by 2045. population policy-2000 propagates target free approach for the family welfare programme in place of coercion. On the other hand it gives informed choice to the people to voluntarily avail the reproductive healthcare services. This approach can be greatly helped by improving the level of literacy which is presently 64.84 per cent. The literacy in girls is further down to 53.67 per cent and it is this section of the population which needs special attention for the success of the population policy. It is pertinent to note that in the present scenario, the sway on the minds of the people on account of social customs and religious faiths has started dwindling with increasing literacy and awareness about the advantages of adopting small family norm. This is an encouraging development even though the results in the rural area are slow in response.

However many people hold a view that a voluntary approach in controlling a birth rate will not succeed and it requires strict implementation of welfare schemes based on two-child norm.

It is interesting to note that there is a geographical divide in India in terms of demography. While there has been decline in the birth rate of Southern States including Maharashtra to come closure to achieving replacement level in not too distant a time, there is no improvement in terms of population control as of now in major states in the northern part viz Bihar, Madhya Pradesh, Rajasthan, UP etc. However there are variations in some parts of the country to the aforesaid generalised statement.

Figure 4.1 shows birth Rate, death Rate and infant mortality rate (urban & rural) from 1991 to 2001.

Although there is a gradual decline in birth rate, death rate and infant mortality rate over the years, all of these are higher in rural area as compared to urban area. The declining trend in birth rate, death rate and the mortality rate indicate that there is a gradual improvement from the past in the level of availability of health services, literacy and standard of living.

The population policy envisages stabilisation of population by 2045. As mentioned earlier, there is a good response to family planning in urban area. However, there is relatively less response in rural area due to high illiteracy and socio-economic reasons. In view of this and in absence of commitment of adequate resources for implementation of population policy, it is unlikely that the population will stabilise in the first half of the 21st century.

2.2. Gross Domestic Product (GDP)/Gross National Product (GNP)

There is a relation between economic status of the country and the allocation for the healthcare. Better the economic status, more the funds are available for the healthcare. GDP and GNP are indicators of the economic status of the country. GNP is a sum of GDP plus net income received from the other countries and thus GNP gives economic status in totality. Per capita GNP represents GNP divided by the total population. India's per capita GDP (2002) is US$ 487 which is much below than even south Asian countries i.e. Indonesia, Sri Lanka, Thailand, China etc. The lower economic status of India is thus one of the major factors affecting adequate allocation to healthcare.

2.3. Income Distribution

The national income if it is not evenly distributed reflect in imbalances in the development in different strata of the society. Against this background poverty is one of the dimensions which affect the living conditions. In India, the concept of poverty line is based upon expenditure required to replenish the energy equivalent of 2100 calories per person per day in rural area and 2400 in the urban area. On this basis 26 per cent of the total population are living "Below the Poverty Line" out of which more than 75 per cent live in rural area.

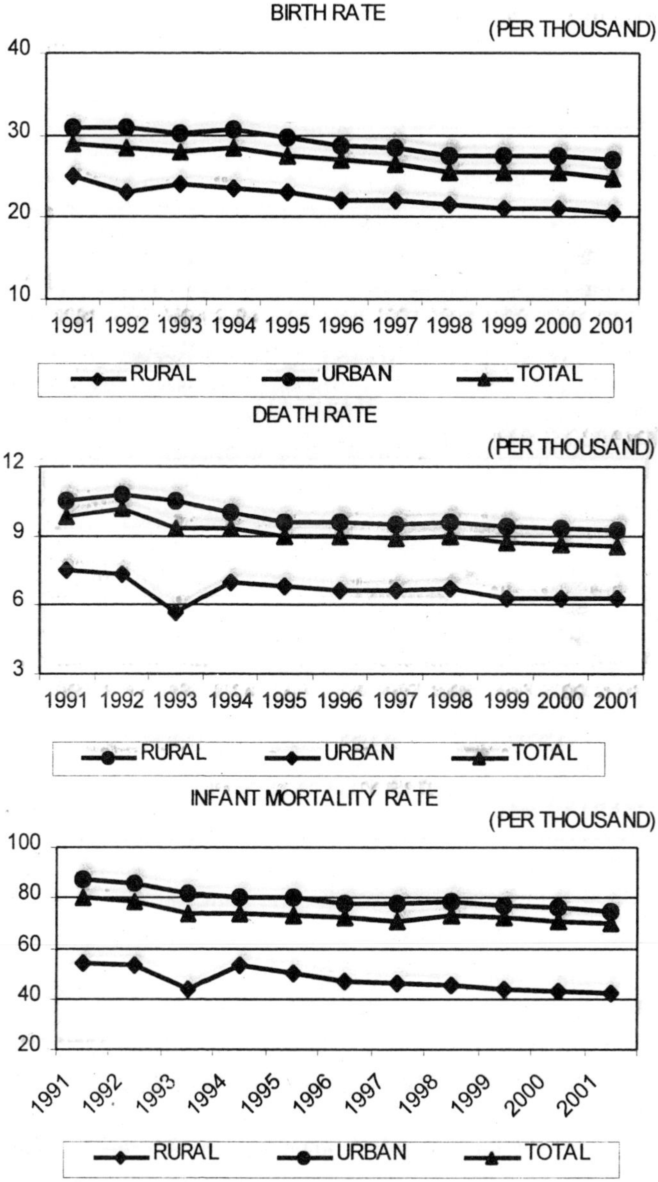

Source: Statistical Abstract India 2003 (GOI Publication)

Fig. 4.1: Birth Rate, Death Rate and Infant Mortality Rate (Urban & Rural) from 1991 to 2001

Generally, the economic status of the person has a direct bearing on the living standard and both have bearing on maintenance of the personal healthcare i.e. poor is the person, poorer is the healthcare. It is therefore incumbent on the State to accord priority of allocation to healthcare so that the goal of Health for All is achieved.

2.4. Work Conditions

The work conditions have tremendous influence on the health. The industrial development has brought in its wake great hazards, not only to the people working in the industries but also people in the vicinity. There are several industrial processes which are fraught with dangers to the health of the workers. The processes and the increased level of pollution are the causes for many diseases such as asthma, bronchitis, pulmonary diseases, heart problems, cancer, deftness etc.

Continuous and three shift working in offices and factories have lead to change in the biological clock for those who are working in these places. These people usually complain of stomach problems and reduced immunity against diseases.

The computers have revolutionised the working in the offices and the factories and made life easier. However the people working continuously on these machines usually complain about backpain, eyestrain and headache.

In rural area it is usually the exposure to hay or the grain dust, contact with the animals and their wastes, exposure to chemicals and fertilizes that cause health problems to the people working in the agriculture. Besides, extreme heat during summer and cold during winter cause health problems in the rural area as they have to work in unprotected and open atmosphere.

Remedial/Preventive Measures: The remedial measures for the above health hazards include:

(i) Using protective masks and devices to avoid inhaling of undesirable gases and particles.

(ii) Rotation of the workers after a period of time and posting them to safer places so that they get healing time for any injury which might cause to their health.

(iii) There should be regular and periodic health check up for the people working in the healthwise undesirable conditions.

(iv) Sanitation facilities such as toilets, proper dwellings, food with required nutrition, safe drinking water etc. will reduce impact of health problems considerably. These have more significance in rural area because of their absence in large measure over here.

Education about Health Hazards: This is by far the most important factor in attaining the best results in the field or healthcare. Every individual has to be made aware of the likely health problems involved and the protective measures to be taken while at work with which he is associated. He would then be in a much better position to avoid getting affected by the health problems. In fact preventive measures mentioned above will reduce disease burden considerably thus resulting in improved productivity and less expenditure on curative treatment of the diseases. A proverb "prevention is better than cure' is more applicable over here.

2.5. Adult Literacy

The objective is to achieve literacy for all persons in the productive age group of 15-35 years. The underlying idea is, if adults are literate, they would be in a better position to understand the importance of healthcare and take advantage of the various schemes being implemented by the State. The literacy in women in particular will make a lot of difference to planning a family. The success of this programme is however dependent upon motivation of the illiterate adults as well as finding suitable time as most of them are engaged in agriculture for earning bread. The teachers would have to follow the timings which are suitable to the adults. Apart from the State, the literate persons in the community, panchayat institutions and social organizations can play important role in educating the adults and continuing education. In this context if one literate person takes a pledge of educating at least one illiterate person, the problem can be easily solved.

2.6. Housing

Lionel Curtis described our villages as dung-heaps. The construction of houses in the rural area has been done without proper planning with the result, even though the surrounding is filled with fresh air and sunlight, there is no fresh air or sunlight in the dwellings. There is no proper drainage system and very few persons have family toilets. Obviously such a situation is a breeding ground for many diseases.

The objective should be to ensure that every family has to have a housing unit of at least minimum standard. In many cases the house sites are allotted by the State to the weaker sections of the society for this purpose as well as financial assistance is provided to them for construction. However the allocation of funds is utterly inadequate to meet the requirement of identified 20 million new units and renovation of 30 million old units.

2.7. Food Availability

India is producing food more than its need. On the one hand while there are regular instances when some of the produce stored in the warehouses for a long time either get spoiled or eaten by the rats, on the other hand there are families in rural and forest area who are so poor that they are not able to afford to buy the food as required for sustenance of the family. According to estimates, more than 200 million people are living "Below the Poverty Line" in rural area. In India, the concept of poverty line is based upon expenditure required for intake of 2100 calories per person per day in rural area and 2400 in the urban area.

Public Distribution System (PDS) has been established primarily to take care of the affordability of the BPL families. The secondary objective of PDS has been to stabilise the prices of the essential commodities in the open market. However there are lacunae in public distribution system (PDS) because of which food is not reaching to all. There are usual instances of good quality of commodities in fair price shops finding their way to the open markets obviously with the connivance of the administering staff of the PDS. If PDS is streamlined, there will be no cause for starvation anywhere in the country.

With a view of giving employment to unemployed and especially during scarcity periods "Food for Work" programmes are taken up by various State Governments from time to time and food is made available to them through PDS. These programmes have met with great success, silver lining being those who participate in the programme get the employment as well as the much needed food. Importantly this also gives impetus to the development work in the rural area, in many cases even remote, which remain neglected otherwise.

3. Provision of Healthcare

3.1. Availability

When India attained independence there were no healthcare facilities in rural area. The people were entirely dependent on their fate. After independence providing basic healthcare caught attention of the Government.

The ultimate goal of a model State is to provide Health for all and it is recognised that the best way of achieving this is through providing primary health care which can cover entire rural population. Based on recommendations of various committees the following healthcare set up has been provided to serve the rural population:

(i) Primary Health Centres & Sub-Centres: Primary Health Centres and Sub-centres are set up in the rural area to achieve the objective of providing health for all. The basic or essential health care is provided at this level. This is the first level at which the community gets in contact with the health system. As on 31.3.2004, 22842 primary hallth centers and 1,37,311 sub-centres have been set up.

Although Bhore Committee recommended setting up one PHC for 10,000 to 20,000 population, due to paucity of resources presently each PHC covers 30,000 population. Similarly each sub-centre is set up to cover 5,000 population. Each PHC is headed by a senior doctor and in sub-centre there are one male and one female health worker. While function of PHC is to provide preventive and curative healthcare, that of Sub-centre is mother and child healthcare, family welfare and immunisation.

In order to serve as a bridge between the community and the health service, a scheme of Village Health Guides has been started in 1977. The guide, generally female, is selected by and out of the community. The guide is a local person and a part time worker. The guide is given orientation training about primary health care and is provided a kit with basic medicines.

One Dai is provided in each village to assist in deliveries, initial mother and childcare

(ii) Secondary level: District hospitals and community (rural) health centres form the secondary level at which better medical facilities are available. These also serve as the first referral hospitals. As on 31.3.2004, 3043 community health centres are set up. Community health centres, upgraded from PHC and situated at each Development Block, covers the population of about 80,000 – 1,20,000. It is manned by 4 doctors and other medical and paramedical staff.

(iii) Tertiary level: Medical college hospitals and specialised hospitals come under this level. At this level, highly sophisticated diagnostic equipments as well as more experienced specialists are available. These hospitals are expected to deal with more complex medical cases.

Presently in a country as a whole there is a ratio of one doctor to 20,000 population. This is very low compared with even China where it is one doctor to less than 10,000 population. Though nearly 72.22 per cent i.e. almost three-fourth of India's population live in 5.94 lakh villages, less than one third trained doctors practice in rural area. There is reluctance on the part of the qualified doctors and other medical staff to stay in Community Health Centres and PHCs which are all located in rural areas. This is resulting in high absentee rates of doctors and medical staff from their duty places . Table 4.2 shows absentee rates (2003) of primary healthcare staff in some of the major states.

Table 4.2: Absentee Rates (2003) of Primary Healthcare Staff

State	*Absentee Rate %*
Bihar	58
Gujarat	52
Karnataka	43
Orissa	35
Rajasthan	39
Uttar Pradesh	42
West Bengal	43

Source: India-sustaining Reform 2003

The reluctance of medical personnel to stay at the Centres is also stemmed from the fact that these centres are not well equipped and there are no good living facilities in rural area as in cities. The turnover especially of the doctors from these centers is therefore very high. This is a major cause for poor functioning of these centres.

It is noteworthy that while institutional framework as mentioned above has been established at number of places, there are still many parts of rural area wherein health facilities are nowhere to be seen as these facilities are located at a place which is far away from them.

3.2. Accessibility

It becomes utterly difficult to address health problems for the people who are away from the PHC or Sub-Centre. Some of villagers have to trudge a distance of 5 to 20 kms to get the medical facility. Patients in rural area being carried in bullock carts to health centers for urgent treatment is a common scene. The problem of distance is further compounded by the bad condition of roads in the rural area. There is also no reliable and regular public transport arrangement. The villagers cannot also afford to be away from their workplace for long. The situation can improve if the medical officer visits the villages or periodically special health camps are held in the villages.

The patients admitted to the hospitals and health centres are advised to purchase special medicines from outside which becomes difficult for poor patients.

On the economic front there is a clear divide in terms of accessibility between the rich and the poor. While there is a proliferation of fitness clubs and ultramodern facilities which are available to rich in the cities, there are instances when rural poor opt for even selling their organs to repay the debts.

3.3. Utilisation

This aspect indicates the extent of utilisation of the health services *vis-à-vis* requirement of the community. Although the population needing healthcare is very large there is a problem of utilisation of healthcare facilities especially in rural area because of the accessibility problem. The villages situated away from the health centres or sub-centres find it difficult to take patients all the way to centres for treatment. There are shortages of essential medicines and other medical facilities in State run hospitals and health centres which discourage the inflow of patients. There are also instances like costly diagnostic equipments in hospitals when they go out of order, are not repaired in reasonable time. This happens' primarily on account of bureaucratic approach of the staff responsible for this work and secondly on account of paucity of funds.

The State has also provided family welfare services but these services are not utilised to the expected level. This is basically due to lack of awareness in rural population about the advantages of family planning. The awareness can be brought about through literacy drive and effective counseling to the eligible couples.

While there is a lot of scope for improvements as mentioned above in the State run health services, they are still the only ones which the rural population can depend upon as they find these economically affordable and relatively nearer.

3.4. Quality of Care

The quality of healthcare in State hospitals is not up to the mark primarily due to paucity of funds and heavy rush of patients. The waiting period for showing the patient to the doctor and the treatment is quite long. The diagnostic facilities are overloaded and there is always shortage of beds. The patients are advised to purchase special medicines from outside which becomes difficult for poor patients. On the other hand, although personalised care is available in private hospitals, it is beyond the means of the rural poor.

Due to health centres not being well equipped and these being located in rural environment accompanied by obvious lack of personal comforts, well experienced doctors are not generally willing to serve at the centres for a long period and there is heavy turnover of the doctors from these centres. This has considerable bearing on the quality of services being rendered by the health centres. The position is so precarious that the admission to medical colleges has been made subject to executing a bond by the students to the effect that they would have to serve the State hospitals and the primary health centres for two years after passing the final examination. No doubt such a compulsion is a necessity for running the health services but the doctors are hardly motivated to stay at the health centres.

Another reason for low quality of care is due to typical bureaucratic approach of the health administrative offices in appointing doctors and procurement of quality medicines in time. It is unfortunate that the efforts of the hospitals and the centres are not adequately backed by their supporting offices.

4 Health Status

4.1. Low Birth Weight (LBW)

Weight at birth is one of the major indicators of the infant's health, chances of its survival and development. Low Birth Weight (LBW) has been defined by WHO as weight of the baby less than 2.5 kg at birth. The percentage of LBW babies (infants) in India (1995-2000) is 26 per cent of all live births as against China's 6 per cent. UNICEF monitors the data based on children under 5 years. According to UNDP HDR report 2004, percentage of underweight children under 5 years (1995-2002) is 47 in India as against 11 per cent in China. The causes of LBW are malnutrition, high workload and low health status of the communities especially rural mothers. Low health status of mothers include infections, blood pressure, inadequate interval in the pregnancies etc.

Preventive care: Preventive care includes timely and proper prenatal care and providing nutritional diet to mothers. While they can do light work, they should be kept away from the strenuous work. The target has been set to bring down the present rate of LBW babies to not more than 10 per cent in India as well as in rest of world wherever the rate is higher.

Treatment: For LBW babies, proper neonatal care is essential. In extreme cases intensive care in ICU and incubatory care are needed which are not generally available in rural area. This is one of the major reasons for high infant mortality rate in rural areas.

4.2. Nutritional Status of Children

In our country major problem is poverty. 26 per cent of the population is still below poverty line. The problem for these people is how to get daily two square meals. The nutritional aspect of the food becomes less important to them so long as their bare requirement of food is not met. It is not that our country is not producing enough food required for the entire population. On the other hand it is producing food more than its requirement so much so that India is exporting food items to other countries and a lot of food goes waste in storages and eaten by the rats. Basically the real problem is that Public Distribution System (PDS) is not functioning properly and requires to be streamlined.

In India, generally predominant factors in selection of food are: food being produced in one's geographical area, age old social customs and habits formed since childhood. It is of late that the importance of nutritional value in the food is somewhat being understood by the people. The awareness of this aspect is still much less in rural area and the nutrition related education through various programmes and mass media can play vital role in improving the present situation.

The nutrients in the food comprise carbohydrates, proteins, fats, minerals and vitamins. The function of the carbohydrates is to supply energy. These are found in cereals, sugars etc. The function of the proteins is body building and its maintenance. These are found in pulses, cereals, oils, nuts, meat eggs etc. The fats provide high calories. The sources are oils-groundnut, coconut etc; ghee, milk; meat etc. Vitamins are organic compounds. Their basic function is not to provide energy but to supply essential nutrients required for maintenance of health. The diet should ideally be based upon the requirement of the total calories and further the content of the nutrients in the food intake. Although certain changes can be suggested in the food pattern of the community or the family, the selection of the food has ultimately to be based upon individual requirement.

It is a paradox that while poor suffer from deficiency of nutritious food, developed countries and rich people in developing countries suffer from obesity which is directly a result of overeacting and overnutrition. This has led to a mushroom growth of slimming clinics all over the world and they are doing roaring business.

4.3. Infant Mortality

Infant means a child with age less than 1 year. The Infant Mortality Rate (IMR) of the country is an indicator of the social and health status of its women and children. At the ground level it reflects the effectiveness of the health services, IMR is therefore regarded as one the most important indicators of the development of the country.

The infant mortality rate is calculated as below:

$$\text{IMR} = \frac{\begin{array}{c}\text{Number of deaths of infants}\\ \text{(children less than 1 year age in a year)}\end{array}}{\text{Number of live births in the same year}} \times 1000$$

With sustained effort worldwide, there has been steady decline in the IMR, more predominantly in developed countries. In India it is 67 (year-2002) as against less than 10 in developed countries. Of this, neonatal (children under 28 days of age) deaths account for 70 per cent of infant mortality. In our country itself IMR varies from state to state and rural to urban population. It is less where education and living standard are high. The IMR is more in rural population as compared to urban population. The causes of the high IMR and remedies for the same are (Park-2005):

- Living Standard: There is a strong relationship between poverty and infant mortality. Poor people are not able to afford proper nutritional food and the hygienic and restful conditions for the expectant mothers which has a bearing on the health of the baby. Balanced economic development across the communities, whether rural or urban, can only remedy the situation.
- Availability of proper Healthcare: While every mother is expected to receive healthcare after conception is confirmed, only 43 per cent (2002) of the deliveries in India are attended by the trained personnel. In rural health centres, facilities like intensive/neonatal care unit to handle complicated cases are not available. The allocation of adequate funds for rural health centres including PHCs and Sub-Centres and its efficient utilisation can enhance the availability of proper healthcare in the rural area.
- Low Birth Weight: This is a major cause of infant mortality. The LBW itself is caused by low living standard of the family which prohibits special care of the expectant mothers, unavailability of pre-neonatal and post-neonatal healthcare and congenital defects. This again is related to the basic problem of poverty.
- Immunisation: There are certain infectious diseases like tetanus, diphtheria, measles, tuberculosis, pneumonia, polio which can be prevented by active immunization. It is essential to cover every infant or child by the immunisation schedule and this is possible only with the cooperation of the parents and the effectiveness of the health services. While the success of pulse polio vaccination programme has been phenomenal, the immunisation levels for other diseases covered under various programmes have not shown satisfactory results.
- Low Birth Spacing: Low birth spacing and frequent deliveries drain the energy of the mothers and this escalates the mortality of the children after the first delivery. There should be a spacing of at least 3 to 5 years and in the existing

scenario, it is prudent to adopt a small family norm. Effective family welfare programme can reduce such mortalities.

- Age of the Mother: The probability of infant mortality is more if the mother is either too young or too old to bear a healthy child. The age of the women in twenties is better for bearing the child.
- Breast feeding: It has been established that malnutrition of the infant can be better taken care of by breast feeding of baby. This also enhances immunity level of the infant against various diseases.
- Hygiene: In rural area the houses are quite old and they have not been constructed with the idea of having good sunlight and fresh air. The toilets are also not available in most of the rural houses. Good sanitary and hygienic conditions can well contribute to reduce the infant mortality.
- Education: The awareness of the causes and remedies among the community, especially in women, coupled with effective family welfare programme can improve the situation drastically.

4.4. Child Mortality

Child Mortality Rate (CMR) relates to children between the age group 1-4 years, thus it excludes Infant Mortality Rate. CMR to a great extent represents true socio-economic condition of the community. CMR is calculated as:

$$\text{CMR} = \frac{\text{Number of deaths of the children aged 1-4 years during a year}}{\text{Number of children aged 1-4 years at mid point of the year}} \times 1000$$

In India, the mortality of the children aged 1-4 years is 112 per thousand (1991). UNICEF monitors the data based on children under 5 years. According to UNDP HDR report 2004, percentage of CMR under 5 years is 93 per thousand live births in India as against 39 in China. The deaths occur mostly on account of diarrhea, pneumonia, jaundice etc The underlying causes are again the same as those responsible for infant mortality except care of the mother in this case.

4.5. Life Expectancy at Birth

Life Expectancy at a given age is the average number of years which a person of that age may expect to live, according to the mortality pattern prevalent in that country. This is considered as one of the best indicators of country's level of development and overall health status of its population. Table 4.3 shows decadal expectation of life at birth in India from 1901 to 2001 and Table 4.4 shows the expectation of life at birth in selected countries (2002).

Table 4.3: Expectation of Life at birth-India 2002

Year	*Males*	*Females*
1901	23.63	23.96
1911	22.59	23.31
1921	19.42	20.91
1931	26.91	26.56
1941	32.09	31.37
1951	32.45	31.66
1961	41.89	40.55
1971	46.40	44.70
1981	50.90	50.00
1991	58.10	59.10
2000	62.80	63.80
2001	62.80	64.00

Source: DGHS (GOI)

Table 4.4: Expectation of life at birth in selected countries 2002

Countries	*Yearns*
India	63.7
UK	78.1
USA	77.0
Japan	81.5
China	70.9

Source: UNDP HDR (2004)

There is a gradual increase in the life expectancy in all the countries. This indicates improvement in healthcare services in general. The Life Expectancy at birth in India in the year 2002 is 63.7. It is much lower than the expectation of life in developed countries and it is even lower than China's 70.9. Japan is having the highest life expectancy. Another significant point is that in most countries, the life expectancy in the case of female is more than the male, as if it is a biological factor or is it that the women take life in its own stride better than the men?

One of the goals of the Health for All -2000 has been the attainment of a life expectancy of 64 years at birth for both males and females.

4.6. Maternal Mortality

Maternal Mortality is defined by the World Health Organisation in simple terms as "death of a woman while pregnant or within 42 days of termination of pregnancy, from any cause related to or aggravated by the pregnancy or its management." Maternal Mortality Rate is calculated as:

$$\text{MMR} = \frac{\text{Number of female deaths from causes related to pregnancy, while pregnant or withing 42 days of termination of pregnancy in a year}}{\text{Number of live births in the same year}} \times 1000$$

MMR in India (2000) is 540 per 100,000 as against 56 in China and less than 10 in European countries and USA. Maternal Mortality is again related to socio-economic status, level of education especially in females and healthcare status of the community. Child bearing being a natural process, most of the maternal deaths can be avoided except very complicated cases. However the biggest problem is that most of the deaths occur in the pregnancy cases which have not been registered with the health services for advice, follow up and treatment. There is a preponderance of such cases when deliveries have not taken place in health institutions or are not even attended by the trained personnel. Mostly these happen in villages where trained *Dais* are not available. The other factors which have influence on the maternal mortality are: (i) ripe age of the woman bearing the child i.e. neither the female should be too young or too old. Age in twenties is quite good for the pregnancy, (ii) Frequent deliveries or too less spacing between deliveries are also causes for mortality, (iii) She should be provided with nutritional diet to avoid incidence of Anemia which is a major cause of mortality, (iv) While light duty will be helpful, she should he kept away from the heavy work. (v) She should be registered with the nearest health centre and should get a card for noting down the health status. There should be prior to delivery at least three visits at regular intervals at the nearest health service centre. This would help in finding out if she has any health problem either related to pregnancy or any other, (vi) In some cases there is a need to confirm if the baby is lying in a proper position or if its development is proper or not. However this can only be done where radiology facility is available. Rural area is at a great disadvantage in this respect. (vii) The delivery should be done only in the health institution or at least by a trained Dai. (viii) Hygienic condition is essential to prevent infections to mother or child, (ix) There should be some way to get emergency assistance if the situation so arises.

It will be observed from the above that maternal mortality is a larger question which has to be addressed in the context of socio-economic development, level of education and the most important–the reach of health services in the rural area.

4.7. Disease Specific Mortality

The disease burden vary from time to time. Due to immunisation programme, the disease burden on account of infectious diseases has declined. On the other hand there is more incidence of diseases like heart and cancer. The heart disease is associated with present day tensions and fast life style and cancer is associated with hazardous environment and intake of toxic substances. A new disease i.e. AIDS has come on the scene in 1985 and it has already spread in most of the countries at alarming rate. SARS is the latest arrival. The incidence of diabetes is widespread but the patient dies from some other disease which is propelled by diabetes. In order to assess the prevalence of a particular major disease and its impact, it helps to find out the disease specific mortality and this data can be very useful in framing the health policy, structuring the health services and monitoring the results.

4.8. Morbidity-Incidence and Prevalence

While mortality is the last post in the human life and is countable in terms of cold statistics, it does not reflect the status of the community's health fully. For example, it does not give the impact of diseases from which the community is suffering. Such a status can be known from the morbidity indicator and this can be inferred from the data like: mandays lost due to absence from work place and absence of students and staff of educational institutions on medical ground; attendance at OPD and admission and discharge from the hospitals etc.

4.9. Disability Prevalence

As already mentioned in morbidity indicator, mortality rate does not represent the community's health status fully. It is in this context that disability rate can be helpful. Disability can either be due to illness, or injury due to accidents. If the disability is such that one gets bedridden or one is unable to perform the duty adequately, and such a duration is deducted from the expectation of life, what we get is the expectation of life free of disability also known as Sullivan's Index. It is a quite difficult task to compile the data to arrive at both morbidity indicator and disability rate. However whatever data can be collected, the same can be used to make the health policy broad based.

6. DEMOGRAPHIC HEALTH OVERVIEW

In India, during the last few decades there has been steadily increase in the expectation of life at birth and decline in the death rate. The Table 4.5 shows the estimated age-specific death rate by gender and place of residence (1999) in India. The life expectancy has increased from 41(1961) to 63.7 (2002). The death rate has steadily declined from 21 (1965) to 8.1 (2002). This is indicative of the improvement in the healthcare services as well as literacy and living standard of the people. However there are certain aspects which are significant in this regard:

1. The life expectancy at birth in India is still lower than developed countries. So also the death rate in India is higher than the developed countries.
2. In India the life expectancy in rural area is less than the urban area. So also death rate in rural area is more than the urban area.
3. The death rate is the highest in the age group of 0-4 years.
4. The crude death rate of females is lower than males.

Table 4.5: Estimated Age-Specific Death Rate per 1000 in India (1999) by Gender and Place of residence

Age Group	*Rural*			*Urban*			*Combined*		
	Total	*Male*	*Female*	*Total*	*Male*	*Female*	*Total*	*Male*	*Femal*
0-4	22.9	21.9	23.9	11.7	12.2	11.2	20.4	19.8	21.1
5-9	2.1	1.8	2.4	1.0	0.9	1.1	1.8	1.6	2.1
10-14	1.4	1.3	1.5	0.7	0.7	0.7	1.2	1.2	1.3
15-19	2.2	1.9	2.4	1.2	1.1	1.3	1.9	1.7	2.1
20-24	2.9	2.7	3.1	1.8	1.3	2.2	2.6	2.3	2.8
25-29	2.9	2.9	3.0	2.6	2.9	2.4	2.9	2.9	2.8
30-34	3.3	3.7	3.0	2.4	3.0	1.9	3.1	3.5	2.7
35-39	4.0	4.8	3.0	3.4	4.5	2.3	3.8	4.8	2.8
40-44	5.3	6.3	4.3	4.1	5.0	3.1	5.0	5.9	4.0
45-49	7.3	8.8	5.6	6.4	7.9	4.5	7.0	8.5	5.3
50-54	11.0	13.2	8.8	8.1	9.3	6.8	10.2	12.1	8.3
55-59	16.8	19.6	14.0	14.8	19.0	9.9	16.3	19.4	13.0
60-64	24.7	28.9	20.8	17.2	21.0	13.3	22.9	27.0	19.1
65-69	37.9	41.5	34.4	33.1	38.4	28.1	36.8	40.8	33.0
70-74	58.7	62.4	55.4	47.9	49.9	46.0	56.2	59.5	33.0
75-79	83.0	92.3	74.2	64.4	66.5	62.6	78.6	86.3	71.3
80-84	107.2	116.1	98.9	90.4	87.8	92.7	103.2	109.4	97.4
85+	167.4	175.8	159.6	160.9	158.2	163.0	165.8	171.8	160.5
All Ages (Crude Death Rate)	9.4	9.7	9.1	6.3	6.7	5.7	8.7	8.9	8.2

Source: Park 2005

It means the situation calls for more improvement in the standard of healthcare services in general and rural area in particular. The major health problem in India is that of communicable diseases. The prominent among the communicable Diseases are:

(i) Malaria: Although the malaria cases have dropped substantially up to 1984, there has been no improvement thereafter; (ii) Tuberculosis: It is estimated that about 30 percent of the total population is tuberculin positive and nearly 5 lakh people die every year on account of this disease; (iii) Diarrhoea: Diarrhoea is a major health problem especially in the children below five years of age. The worst form of this disease can be epidemic of cholera. About 7 lakh people die on account of severe diarrhoea; (iv) Acute Respiratory infection: This disease is a major cause of mortality especially of the children below five years of age. 9.8 lakh deaths were reported in 1998 due to this disease; (v) Leprosy: The prevalence rate of leprosy is about 3.73 percent per 10,000 population. The proportion of infectious cases varies from 6-8 percent. India accounts for nearly 60 per cent of the leprosy cases in the world; (vi) Filaria: There is an increasing trend of the incidence of this disease and it is estimated that nearly 40 percent of the total population is at risk from this disease; (vii) AIDS: After the detection of first AIDS case in 1986, there is a steady increase in the number of cases of this disease. The cumulative cases have risen to about 40,000 and 5.1 million HIV positive cases reported so far.

7. HIPPOCRATIC OATH FOR MEDICAL PROFESSIONALS

Every medical professional is supposed to be under Hippocratic oath. Like Commandments for pious living, Hippocratic Oath states ethics to be followed by the medical professionals. Among other things it states, "I will use treatment to help the sick according to my ability and judgement, but I will never use it to injure or wrong them". However, in practice, as in other fields, a lot of commercialization has taken place in medical profession. When people find difficulty in quality as well as easy availability in Public Health Service, some of them turn to private health practitioners. However this treatment has of late become quite costly and beyond the reach of general public. The main factors contributing to high cost of the treatment are:

Doctors tend to prescribe medicines as dictated by the pharmaceutical firms and in return doctors are routinely wooed with attractive packages. The medicines thus prescribed are generally new and costly. There is understanding between the general practitioners and the specialists and referral in these cases result in more expenditure on the part of a patient. Likewise doctors tend to suggest more tests for arriving at the exact treatment and the result is that there is a mushroom growth of diagnostic centres. Doctors however answer this phenomenon by pointing out their fingers towards likely harassment by patients at the hands of Consumer Courts.

Notwithstanding above, there are also many honest doctors who follow Hippocratic Oath but they have to struggle a lot to become popular. To make medical treatment friendly to both patients and doctors there is a need to bring general awareness both in patients and doctors about the ill effects of the above practices and strive towards attaining the ethics propounded in the Hippocratic Oath.

8. WORLD HEALTH ORGANISATION REPORT-2003

The need to equitable distribution of healthcare and allocation of resources for the same in our country has been highlighted earlier. However WHO Report 2003 goes further and refers to the stark disparities between the rich and the poor nations. As an example it has pointed out that while a girl child born in Japan receive medical support on an average worth $ 550 per year, a girl in Sierra Leone does not get more than $3. It highlights that the key task of the global community is to close the gap between such contrasting lives. It calls for delivering quality services equitably and efficiently.

According to the report, among the leading communicable and non-communicable causes of death and disability, HIV/AIDS emerges as the most urgent.

There is a need to have greater interdependence to fight the outbreak of hyper communicable diseases like AIDS and SARS.

9. CASE STUDY OF HEALTHCARE FACILITY IN A VILLAGE-NATALA

The village Natala does not have a primary health centre or a sub-centre. Nearest primary health center is more than 5 km away. There is however one health worker in the village.

The primary health centre functions with one MBBS doctor and one BAMS doctor. The doctors do not continue more than 2-3 years in the health centre. There is a shortage of equipment in the centre and whatever are there, some of these are not in the working order One of the doctors and some of the para-medical staff do not stay at the PHC and their availability in the centre is always uncertain. PHC does not have ambulance but it has a jeep. If it goes for repair or diesel is not available, even emergencies cannot be met. There are instances when patients had to be brought either on bicycle or bullock cart from their villages to the health centre.

A new rural hospital situated 25 km away from the village is looked after with only one doctor. The building for this hospital has been constructed by the world bank aid. Although a male doctor and a female doctor were appointed at the hospital a few days ago, they tendered their resignations. Over 300 patients visit OPD of the hospital every day. In addition, 20-25 patients have been admitted into the hospital. Obviously it is extremely difficult to manage entire hospital by one doctor alone.

Villagers are unable to face emergencies. During rainy season it is all the more difficult to get the medical aid as the *nallah,* which is passing nearby, prevents villagers to reach the nearest health centre. There were instances when timely aid could have saved life of some of the villagers. The villagers are so scared of the illnesses that they always pray God for their well being rather than anything else.

The doctor visits the Aaganwadi once a month. The villagers are very happy when the health camp is held in the village. The last health camp was however organized in this village 8 years back. During the camp the doctors and specialists come from district level hospitals. The response of the villagers to the camp is simply overwhelming.

There is no underground drainage system or public lavatory. There are only open drains. These are usually overflowing and filth in these drains cause health problems to the residents. The human encroachments have made the drainage system worse. Only few families have proper toilets. A scheme of subsidised toilets was introduced in the village according to which for a toilet costing Rs. 2000/-, only Rs. 500/- was to be contributed by the resident. Some of the villagers who opted for the scheme are using the constructed toilet as a bath room obviously because they do not have any bathroom especially otherwise required for the ladies.

10. STRATEGIES TO BE ADOPTED TO ACHIEVE A GOAL OF HEALTH FOR ALL

While the achievements due to the efforts of the Govt. and Non-Govt. organisations in improving the healthcare are not small, the fact remains that the goal of achieving Health for All is still far away. It would be impossible to achieve the goal of Health for All, which is by no means easy, without adopting necessary strategies. These strategies have been mentioned above. However main strategies are discussed below:

1. Striving for Community Participation

The rural masses have to be educated more about the preventive aspect of the health such as maintenance of hygiene, sanitation and drinking safe water. The last posts in rural area i.e. *Dai* or *Health guide* can play vital role in generating awareness in the local community about the general approach towards health problems. A message has to go to the community that they have themselves a vital role to play in promotion of their own health. The infrastructure for this like sanitation, clean drinking water etc. will have to be provided obviously by the Govt.

As regards curative aspect, holding periodical medical camps in rural area has been found to be quite effective. There is a need to cover even remotest village in the medical camps. For involvement of the community, enough publicity prior to the date of holding the camp is necessary. Also association of NGOs can be very useful to make the medical camps successful.

2. Improvement in Organisation Structure

Presently the medical staff in the rural health centres are coming from two streams: one stream comprises staff appointed by the State Govt. and the other appointed by the Zilla Parishad. This is creating dualism in the system and ways have to be found to avoid the conflict arising out of the dual system.

In the present set up there is a great scope for political class and the bureaucracy to play a role which many a times becomes a hindrance for proper functioning of the health centres. In this context decentralisation of power by the State to Panchayat Raj Institution will make a lot of difference. The improvement is already perceptible in the areas in which powers have been delegated to PRIs.

3. Easy Accessibility To Health Services

In rural area, main hindrance in availing the heath services is the problem of accessibility. Apart from the distance at which the health centres are located, there is a serious problem of bad condition of roads. The accessibility problem can be tackled by: (i) Locating health centres nearer to the remotest village say within a distance of 3 to 5 km, (ii) improving the condition of roads connecting the villages to centres, (iii) Improving the public transport system on these roads and (iv) periodical visits of the medical staff to villages and arranging medical camps in villages.

The institution of *barefoot* doctors that started in China to cope up with the vast rural population yielded excellent results. India too is facing similar problems and therefore that system should equally work well over here. In India already health guides have been employed to work as a link between the community and health centres.

4. Imrovements in Functioning of Primary Health Centres & Sub-Centres

At present the functioning of Primary Health Centres & Sub-Centres leave much to be desired. The reasons for such a state of affairs are found to be internal as well as external. Among the internal factors, the most important is the attitudinal aspect. The disposition of the staff working in the centres has to be sympathetic towards the problems of the rural people who are not generally literate and it may not be out of place to mention that the motto of the staff should be to serve the underserved. Sensitising the staff through training at the time of induction as well as refresher courses during service can be very useful for this purpose.

The external factors are: At times the role of bureaucracy is not supportive to proper functioning of the centres. Since the general public have certain expectations from the centres with which they are directly interacting, the political class and the bureaucracy should extend all necessary support without which the centres cannot function properly.

The other area of concern is that the health centres are not adequately equipped. This became quite apparent in a 15-day training programme held in Maharashtra where 100 rural doctors participated. The programme was organised for rural doctors as part of a World Bank Project. Rural doctors expressed that their hospitals are not having endoscopy equipment, while 50 percent of the gastro-intestinal emergency cases end in death. Professionally, doctors feel disconnected from the latest progress taking place in the diagnosis and the treatment of the patients. Even medicines essentially required are not provided in time. Proper residential facilities for the medical staff are not available; Educational facilities for their children are not available as good as in urban area; In nutshell there is a clear Rural-Urban divide in terms of availability of facilities both in centres as well as on residential front. The rural doctors always compare with their counterparts in towns and cities and the rural posting is taken by them as a transition for a change to urban location. This has lead to concentration of doctors (upto 73.6%) in urban area where only 27.78 per cent of population lives. The only way to improve this situation is to provide necessary facilities at the centres so that the doctors and the medical staff are motivated to work with devotion and efficiency.

The punctuality in attendance of the doctors and the medical staff is another problem area whenever these persons live at a place different from the place of posting This affects the working of the rural health centres. It would be much better if the facilities are developed at the centres to motivate them to stay at the centres. In spite of providing adequate facilities if there is no punctuality, strict disciplinary action is the only alternative. However since most of the staff does not maintain punctuality, in the first place, no disciplinary action is initiated against the erring official. If there still happens to be any case of disciplinary action, the erring official invariably approaches somebody in the political class or bureaucracy and gets the action scuttled.

It is therefore necessary that there should be no interference from outside when there is a question of taking disciplinary action against an erring official.

5. Population Strategy

The international study report compiled by exports from UN agencies and international scientific and development organisations spread over 95 nations and released on 30.03.2005 by the 45-member board of the Millennium Ecosystem Assessment says, "The human beings are damaging the planet earth at an unprecedented rate and raising the risk of abrupt collapses in nature that could spur disease, deforestation or 'dead zones' in the seas. Rising human population had polluted or over-exploited two-thirds of the ecological systems on which life depends, ranging from clean air to fresh water, in the past 50 years".

Figure 4.2 shows the population figures of top 5 countries of the world, in which India stands second to China. Presently world's population is estimated to be 6.3 billion plus. The population of India as per 2001 Census is 1028.7 million. The population at India and China together constitute 38 per cent i.e. more than one-third population of the world. Further population of five countries i.e. India, China, USA, Indonesia and Brazil accounts for almost half of the world's population. During a decade 1990-2000, it grew at an annual rate of 1.4 per cent. The annual population growth rate during the same period in selected countries is:

Country	*Annual Growth Rate (per cent)*
World	1.4
India	1.9
China	1.0
USA	0.9

China has achieved almost the same annual growth rate as USA.

The population of India is 16.87 per cent of the world's population as against country's area accounting for 2.42 per cent of the world. In this context, Table 4.6 Shows the status of crude birth rate and death rate (of which population is a function) of some of the selected countries.

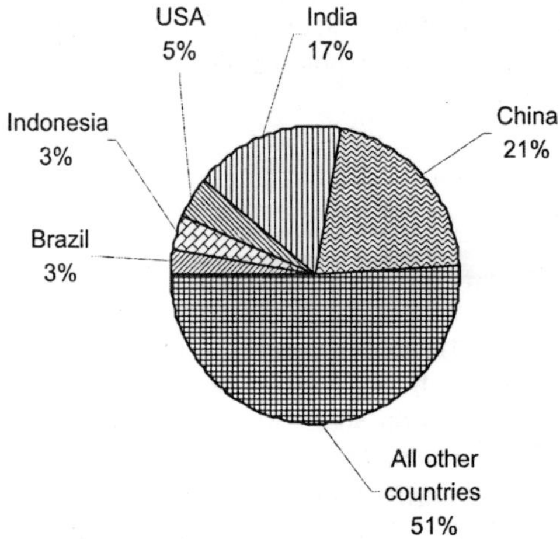

Fig. 4.2: Top 5 Populous Countries of the World-2001

Source: For India: Census 2001, Other Countries: World Population Day Reports

Table 4.6: Crude Birth Rate and Death Rates (per 1000) in Selected Countries in 2002

Country	*Crude Birth Rate*	*Crude Death Rate*
India	25	8.1
China	15	7
Japan	9	8
UK	10	10
USA	15	8

Source: Park-2005

It is seen from the above table that India's crude birth rate is 150 per cent of the second highest that of China.

Table 4.7 shows India's popoulation and its growth by religion during las two decades as per Census 2001, among the major communities namely Hindus and Muslims who constitute 94.1 per cent of the total population, the growth rate in population of these communities has declined in the decade (1991-2001) from the previous decade (1981-1991) by 2.8 and 3.6 per cent respectively. During the decade (1991-2001) itself, the population growth rates in these communities is 20.0 and 29.3 respectively.

Table 4.7: India's Population and its growth in percentages by religion during last two decades.

Religion	*Population (Adjusted) as per census in*		*Growth in population w.r.t. previous decade*		*Increase (+) /decrease (–)*
	1991	*2001*	*1981-91*	*1991-2001*	
All religious communities	100	100	23.8	21.5	(–) 2.3
Hindus	82.4	81.4	22.8	20.0	(–) 2.8
Muslims	11.7	12.4	32.9	29.3	(–) 3.6
Christians	2.3	2.3	17.0	22.1	(+) 5.1
Sikhs	2.0	1.9	25.5	16.9	(–) 8.6
Buddhists	0.8	0.8	36.0	23.2	(–) 12.8
Jains	0.4	0.4	4.0	26.0	(+) 22.0
Others	0.4	0.7	13.2	111.3	(+) 98.1
Religion not stated	Negligible	Negligible			

Source: Census of India 2001; Website: <www.censusindia.net>

The major reason for the relatively higher growth rate in Muslim community vis-à-vis Hindus are cited as:

(i) Education: 65.1 per cent persons above the age of 7 are literate among the Hindus and 59.1 per cent among the Muslims.

(ii) Poverty: The rate of work participation in Hindus is 40.4 per cent and Muslims 29.3 Poverty could be another reason for higher growth rate.

(iii) Socio-cultural attitude of the people

As regards Christian and Jain Communities, the growth rate in the decade (1991-2001) has increased from the previous decade (1981-91) by 5.1 and 22 per cent respectively.

The growth rate of religion-communities in last decade (1991-2001) are Christians (22.1%), Buddhists (23.2%) and Jains (26%).

Any social change to happern takes time. It would require more efforts to bring down the present population growth rate whichever community it relates to i.e. Hindus, Muslims, Christians, Sikhs Buddhists, Jains or any other. Cultural diversity enriches people's lives and therefore, if handled well, religions identity which may result in cultural diversity becomes a positive factor. However, the steady increase in India's population is a matter of serious concern as the availability of resources for development of various sectors including healthcare has a limitation. India's Crude birth rate (2002) is 25/1000 and crude death rate is 8.1/1000. Higher birth rate means overstretching already strained public services. The annual growth rate of population is 1.95 per cent and at this rate population will be doubled within 30 years. As per Health Policy it was estimated that population will stabilise by the year 2045. However as per study on India's population by Prof Tim Dyson of London School of Economics and Political Sciences, India's population is likely to be in the vicinity of 1400 million by 2026 and 1600 million by 2051 before the "steady-state" level is reached by 2051. The projections, however, employ what some would regard as fairly optimistic assumptions about the future demographic progress namely the total fertility rate will fall to around 2.1 per cent births per woman by 2016-21.

China had problem similar to India but they contained it by adopting one child norm The target of India's population growth rate is is set at 1 percent as against the growth present rate of 1.95 percent. India being a democratic country, it is difficult to employ coercive methods for restricting the family size. The family planning programme has therefore been renamed as family welfare programme. However this programme could not make significant impact in rural area especially in tribal area. There is a greater need to spread awareness in these areas about the importance of adopting small family norm and organise effective campaign for adopting two-child norm.

Two Child Norm: some of the persons hold the opinion that application of two-child norm will diminish the women's social power that they possess today or that the norm will hit the poor. They ignore the fact that the woman today is not being respected on account of reproductive power alone but due to her many other qualities like service before self, compassion, feelings etc. Rather more deliveries tell upon her health and become an hindrance in the pursuit of career she wants to choose.

Apex court in its judgement (2003) upheld the Haryana Govt's two-child norm that debards anybody with more than two children standing for election as sarpanchas or upsarpanchas. The Court ruled: In our view, disqualification on the right to contest by having more than two living children does not contravene any fundamental right nor does it cross the limits of reasonability. The ruling, after referring to the demographic growth status in the country appreciated the objective behind the legislation. Maharashtra, Rajasthan and Andhra Pradesh are also contemplating to introduce such a provision.

There is no denyng the fact that in the long run education, development and woman and child welfare are better ways to lower the family size vis-à-vis force, coercion, incentives and punitive disincentives and this particular approach i.e. reproductive health rather than population control was also accepted as a motto at the Cairo Conference in 1994. However due to severe resource constraint as well as socio-cultural ethos, it is difficult to achieve the objective referred above for a long time and therefore present in intervening measure like adoption of two-child norm to bring down the desired population growth rate becomes essential. The policy can be reviewed from time to time and two-child norm can be relaxed/done away with at appropriate time if necessary.

Another problem area is that the female to male ratio is falling which signifies continuation of social bias in favour of male child. As per 2001 Census, on all India basis, the number of females per 1000 males is 933. It is quite interesting to note that the sex ratio in rural area (946) is more favourable to female than in urban area (901). Property is a major cause for skewed sex ratio. The rich want only male heirs to inherit their property. It is unfortunate that contrary to general perception, literacy rate has not made positive impact on correcting this bias. In this context social reformers and institutions can play vital role in correcting the general perception of the people in this regard.

5.1 National Commission on Population (NCP)

The National Commission on Population was constituted in May 2000 to oversee and review the implementation of the NPP. The states have been identified which are contributing to higher growth of population. These are mostly central and northern states viz. Bihar, Madhya Pradesh, Rajasthan, UP, Uttaranchal, Jharkhand, besides Orissa and Chhattisgarh. These states are projected to contribute 55 per cent increase in population during the period 1996-2016. The Empowered Action Group (EAG) has therefore been constituted to facilitate the preparation of area-specific programmes with special focus on these states which have been lagging behind in containing population growth to manageable limits.

National Population Stabilisation Fund (NPSF)/Jansankhya Sthirata Kosh (JSK):

The fund has been registered as a Society in June 2003 with a seed money of Rs. 1000 crore, to provide a window for canalizing funds through various sources to especially aid projects for population stabilization.

6. Promotion of Family Welfare Schemes

India's Population Policy-2000 promotes giving informed choice to the people to voluntarily avail the reproductive healthcare services in place of setting predetermined targets. It is seen that there is unevenness in our country in achievement of population stabilisation programme. While there has been decline in the birth rate of Southern States including Maharashtra which are coming close to achieving replacement level in not too distant a future, there is no significant improvement in terms of population control as of now in major Bihar, Madhya Pradesh, Rajasthan and UP states in the northern part. The Govt.'s decision to do away with sterilization targets has also affected family planning programme in these states. This indicates that the family welfare schemes have been better administered in southern states as compared to northern states, save some exceptions. Northern states have therefore to replicate the efforts made in the southern states. The improvement in the literacy level in rural area has a great relationship with the success of the family planning programme and therefore outreach of literacy in rural area including tribal area especially in girls/women is a key element in achieving objectives of the family welfare programmes. It is unfortunate that in present times while there is emphasis on various methods of family planning including use of contraceptives,

IUD insertions, sterilization, etc., nobody talks about practicing continence i.e. exercising self control, which is much healthier way to achieve the objective of family planning. However contrary to the policy adopted towards family welfare, many people strongly hold a view that a voluntary approach in controlling the birth rate will not succeed and it requires strict implementation of welfare schemes based on two-child norm.

6.1 Reproductive and Child Health (RCH) Programme

The programme (with district as unit) follows a differential strategy with inputs linked to the needs of the area as well as capacity for implementation. It is a flagship programme of the Family Welfare, which combines the objectives, viz. reproductive health, child survival and fertility regulation.

The programme is operated though Primary Health infrastructure. Dai training also forms part of the programme. The overall goals of the programme are to reduce maternal and infant mortality and morbidity and assure reproductive health and choice and thereby contribute to stabilization of population. The programme is supported by World Bank, European Commission, UNICEF and other bilateral donors.

7. Assigning Running of Public Health Centres to Competent NGOs

Presently Primary Health Centres and Sub-centres are functioning under State Health apparatus which experiences difficulties due to bureaucratic approach from above and less accountability of performance within the Centres. Many doctors reside away from the centres and some of them are not punctual in attending to their duty. They also do not visit villages under their charge regularly. One of the ways to make Health Centres more effective is to assign these centres to NGOs who are capable of handling medical services in rural area. On pilot basis a few centres have been assigned by Maharashtra State Govt. to Mahatma Gandhi Medical College at Sewagram and they are performing very well. In this case, the set up and doctors belong to State Govt. and salaries and other expenditure is also born by the State Govt. The administration and functioning of the Centres has been assigned to the NGO.

8. Combating Malnutrition Deaths Protecting Child Rights

In 2004, reports of malnutrition deaths in certain states such as Maharashtra caught the attention of the media even as this phenomenon is prevalent throughout the country.

Child malnutrition contributes to more than 60 per cent of the 2.4 million deaths under-five child deaths every year. Two-third of these deaths occur during the first year. Early solutions are needed, as during first two years malnutrition sets in and is virtually irreversible. Also the development of the brain occurs in the first three years and therefore intervention at this stage becomes crucial.

According to WHO, child malnutrition is related to inappropriate infant feeding practices, which are very poor in India. Exclusive breastfeeding for the first six months is practiced in about 40% of the infants and appropriate complementary feeding after six months in about 33 per cent. Adults succumb to commercial advertisements buying baby foods, which only perpetuate relatively poor feeding. The solution lies in improving infant and young child feeding practices through countrywide breastfeeding support centres in all facilities including public and private ones.

Further it is not adequately understood that at an early age children are dependent on care-givers, whose knowledge about infant and young child nutrition is poor. India's response to the issue takes the form of the Integrated Child Development Services (ICDS) programme. Apart from the fact that ICDS reaches only a quarter of the child population, the programme is heavily dependent on the performance of Anganwadi worker, who does not possess sufficient knowledge and skill. Preoccupied which so many other tasks, she is unable to provide necessary attention to this particular aspect. Family based interventions with community involvement would be more effective in this respect. The high rate of infant mortality, low birth weight, low resistance to infections and vulnerability to various diseases are generally the effects of the malnutrition. The manifestation of malnutrition lead to diseases like protein energy deficiency, ophthalmic problems, anemia, goitre (due to iodine deficiency), heart disease, diabetes etc. India has perhaps highest prevalence of nutritional anemia in women and children. About 50 per cent of non-pregnant women and young children and 50 to 80 per cent of pregnant women suffer from anemia. Also 20 to 40 percent of maternal deaths are attributed to anemia. Recent surveys indicate that in rural India, anemia is much more widespread, even in men.

The malnourishment level in kids is measurable and these are classified as grade one to four (highly malnourished). Those children whose weight falls below fifty percent of normal expected weight are catagorised in grade three and four. It is advisable to admit grade three and four kids in the hospital till their grade improves. However there is great reluctance on the part of some parents even to admit the affected child in the hospital and instead they approach some unprofessional persons or even quacks for cure of the illness

The incidence of malnutrition deaths is more in tribal area. Besides improper feeding poverty, early marriages, superstition and lack of hygiene are some of the major reasons for malnutrition deaths. To get over the problem, there is a need to launch effective awareness campaign especially in tribal area about the risks due to malnutrition. High incidence of malnutrition deaths is also an indicator of poor performance of the Public Distribution System and absence of even minimum primary healthcare in that area.

Considering importance of the issue, it is proposed to introduce a bill seeking to set up a National Commission for the Protection of Child Rights in keeping with the UN Convention on the Rights of the Child (CRC) in 1992. The Commission will oversee the implementation of the National Policy of Children, which is expected to improve the survival rates, health, nutrition, education and over aspects for all children especially girl child

Lastly for the above programme (ICDS etc.) the budgetary provision for under-six children need to be substantially increased from the present allocation of Rs. 3,142 Crore for 2005-06.

9. Revamping Public Distribution System

In a country with surplus production of foodgrains and large bufferstock, 200 million people remaining hungry reflect poorly on the functioning of the Public Distribution System. There is a need to revamp the entire system and then any tampering of the system in place should be dealt with severely. Instead of storing food in centralised distant places, local food banks should be created so that movement and utilisation of foodgrains become easier and cheaper.

According to the survey conducted in 2003-04, the Planning

Commission's programme evaluation division has found that more than 50% of the foodgrain meant to go to people living below the poverty line did not reach them. Every year about half the food subsidy targeted for the poor and worth Rs. 4,000 crore slips away. The main reason for this is the food subsidy creates duel pricing—a subsidised, low price and the market rate for the same thing, which creates an opportunity for the traders to buy cheap and sell dear. One of the ways to plug the leakages is to issue food stamps to eligible poor who can exchange them at any shop.

10. Safe Drinking Water and Sanitation

10.1 Safe Drinking Water

Although many of the villages have been provided with hand pumps for drinking water, there are quite a number of habitation in remote area where people have to fetch water from a distance. Easy access to safe water is a priority area as many of the health problems like infectious diseases are caused by the use of unsafe drinking water.

10.2 Sanitation

Although rural people are quite particular about personal hygiene, they are not so about maintaining hygiene in public places. In rural area, people often go in the open for defecation. However we cannot blame rural people only for this situation. The factors like nonavailability of space in the house for constructing the toilet, unawareness about the simple design of the toilet and poverty very much contributed to the problem. More than anything else, toilet facility is important from the point of view of providing privacy and dignity to women. As per 2001 census only 21.9 per cent of the total 138.3 million rural households are having latrines. Further bare 6 per cent of the house households are having toilet facilities. Most of the State governments have started the scheme of *Sulabh Shouchalaya* in villages, which provide subsidy to the extent of 75 per cent. However the response level to the scheme is still about 15 percent. There is a need to carry out awareness campaign in the rural area about the importance of hygiene/sanitation. This can be best done by the Gram Panchayat/Gram Sabha and NGOs can also play important role in this respect.

The response to the above scheme may also improve if the subsidy is raised from the present level. On the contrary there is a move now to reduce the subsidy and in its place a programme of spreading awareness in the community through schools about the importance of sanitation is being suggested. It is unlikely that the programme-Total Sanitation Campaign (TSC) would succeed only through awareness campaign without offering substantial subsidy.

11. Implementation of Mid-Day Meal Scheme in Schools

No other Govt. sponsored scheme has become so popular as mid-day meal scheme in schools, wherever it is under implementation. This scheme is Centrally sponsored and covers children enrolled in Aanganwadis and Primary classes i.e. up to class V. Prior to this scheme, children from poor families used to avoid going to school because the families could not afford to send meals with the children. The cooked meal under this scheme is supposed to be hygienic and contain energy of 300 calories and up to 12 gm of protein. This scheme has enabled to improve not only enrolment in schools but also nutritional status of the school going children. There is a proposal to extend the scheme up to class X. If it is not feasible to extend the scheme up to class X, it can at least be extended up to upper primary classes i.e. class VIII.

However not all the State Governments have implemented the scheme in letter and spirit. In many parts of the country, free ration in the form of uncooked rice or wheat is distributed instead of cooked meals. This does not serve the intended purpose. From children's health point of view, it is necessary that the scheme as designed should be implemented. The Apex court has already given a deadline of January 2005 for all states to provide cooked meals to students up to class V.

12. Use of High Technology for Providing Specialist Health Services

While at one end of the paradigm there is a resource crunch in providing much needed minimum healthcare services to the rural population, there is a great surge in the growth of hospitals in cities providing high-tech treatment and residence facilities similar to those provided in five star hotels. The treatment in these hospitals can be afforded only by urban rich.

The facilities in these hospitals are also attracting patients from abroad. In USA and UK patients have to wait for weeks and even years for getting the treatment. It is very expensive to get speedy treatment there. In this backdrop, patients from UK find it preferable to get treated in India. They do not find travel to India in any way as inhibitive factor.

As regards making available specialist treatment in rural area, Karnataka is the first state in the country to operationalise the tele-medicine project at the district/taluk level.

Through this project all district hospitals would be hooked up to super-speciality centres. Already the nationwide tele-medicine pilot project has linked 40 odd hospitals in remote areas including Jammu and Kashmir and the North-Eastern states, to 14 super-speciality centres. In order to increase the access, a satellite exclusively dedicated to providing health services to remote locations in the country is being launched. The aim of this scheme is to use high technology to provide health services to people even in the remotest parts of the country.

Besides, with a view to regionalise the medical expertise, it is also envisaged to open five more medical institutions in the country on the pattern of All India Institute of Medical Sciences at Delhi.

13. Unified Approach for Communicable Diseases

While there are many advantages of going global, one of the biggest disadvantages is the deadly communicable diseases can now spread throughout the world with alarming speed. The latest disease to strike globally in 2003 has been identified as Severe Acute Respiratory Syndrome (SARS). The disease has first originated in China's Guangdong province. It quickly spread not only in China but even distant countries like USA, Canada, Spain, Italy, Sweden, Switzerland, South Africa, Kuwait, Russia, Romania, Philippines, Singapore and even India. There were 750 deaths out of the reported 8100 cases. The disease was spreading in different countries with such a menace that all countries introduced screening of passengers at airports for symptoms. Although the disease has been brought under control, its reappearance any time is not ruled out.

Earlier it was Acquired Immunity Disease Syndrome (AIDS) which is called a ticking bomb. It started in 1980's and now there is hardly any country which has been spared by this disease, worst affected being Asian and African countries. In India alone, there are

5.1 million HIV positive cases reported upto the end of the year 2003. Because of the stigma that attaches to the infection, many sufferers refuse to be tested for the virus. The actual number of affected persons would be much more. The disease is mostly (84.3% cases) spreading through sexual transmission. It has affected not only grown up in families but even children. The cost of medicines for AIDS has been very prohibitive. However there has been an agreement between Govt. and drug companies which is expected to lower the cost of anti-retroviral medicines substantially. Bill & Melinda Gates Foundation in USA has offered substantial grant to India to contain the disease. Even General Trade Agreement on Trade and Tariff (GATT) now World Trade Organisation is moving towards free flow of export of medicines for treatment of AIDS as well as Tuberculosis by softening the patent law which calls for export of patented product only after the royalties are paid.

The moot point is highly communicative diseases like AIDS, SARS, Tuberculosis etc have global ramifications and therefore require unified approach by all nations irrespective of its being rich or poor. Internally, in our country people especially rural population have to be well informed about the preventive measures, safe practices and available health services for any assistance required.

14. Providing Hospital on Wheels

For a spread out country like India, it is necessary to consider innovative means to take primary healthcare to remote villages in rural area. In remote places and hills, transporting patients from one place to another is cumbersome and expensive. In this context, it is worthwhile to consider deploying mobile vans equipped with medical facilities on board. Such vans are already being operated by some of the major hospitals in India with success. To make the scheme effective, the programme of visit by the medical team has to be made known to the villagers in advance. These vans can play a great role in extending the health outreach provided they are monitored and supervised by a dedicated implementing agency. To avoid red tape usually experienced in Govt. working, one of the options is to involve an NGO as an implementing or nodal agency and doctors can be assigned by the health department to them as on deputation.

The cost of each van will be in the vicinity of Rs. 30 to 50 lakhs. Apart from Govt., big corporates can contribute liberally in this venture.

15. Establishing More Medical and Nurses' Colleges

The Planning Commission's latest review of the country's health sector shows that there is just one doctor for 1,800 people and one bed for 1,123 people. In China, with which India is often compared, there is one doctor for 850 people. Govt. hospitals are therefore in desperate need of doctors and paramedics.

There is heavy shortage of doctors and para-medical staff particularly at rural health centres. Table 4.8 shows the manpower ststus in rural primary health institutions in 2000.

Table 4.8: Manpower Status in Rural Primary Healthcare Institutions-2000

Category of Manpower	*No. of Manpower*	*Requirement (base–Census 1991)*	*In position as on 30.6.2000*	*Shortage*	
				No.	*%*
Specialists	4/CHC	22348	3741	18607	83
Doctors at PHCs	1/PHC	22349	25506	3157(s)	14(s)
Block Extension/ Health Educator	1/PHC	22349	5508	16841	75
Pharmacist	1/CHC+1/PHC	27936	21077	6859	25
Lab. Technician	1/CHC+1/PHC	27936	12709	15227	55
X-Ray Tech./ Radiographer	1/CHC	5587	1768	3819	68
Radiographer	1/CHC	5587	1768	3819	68
Nurse Midwife	7/CHC+1/PHC	61458	17673	43785	71
Health Asstt. (M)	1/PHC	22349	22265	84	Neg.
Heath Asstt. (FM)	1/PHC	22349	19426	2923	13
Health Worker (M)	1/SC	134108	73327	60781	45
Health Worker (FM)	1/SC + 1/PHC	156457	134086	22371	14
Total		525226	337086	188140	36

Source: NIRD Rural Development Statistics, 2002-03 (s) Surplus

In all categories of doctors and medical staff there are shortages except doctors at PHCs. Even in PHCs doctors do not continue for long.

In 1980's, apart from Govt. collages, there were less than 10 private medical colleges in the country. On account of Government's inability to open more number of medical colleges due to financial constraints, in 1990s Govt. allowed private initiative in opening of more and more medical colleges. Thus private medical colleges spruced up in a big way. It was then found that not all organisations had capability of providing minimum facilities required for opening medical colleges and hospitals. It was therefore decided that opening of any new medical college will be subject to satisfaction of MCI/ GOI (Medical Council of India/Government of India) about such a capability. As a consequence of this, pace of opening of fresh medical colleges considerably slowed down. Due to inability of Govt. to set up new medical colleges and the stringent norms applied by the MCI/ GOI for opening of new medical colleges even by the private organisation, the result is, today we find tremendous gap between the requirement of doctors and the number of doctors coming out of the medical colleges. Besides, there are serious problems about the mode of admission to be adopted and the fee structure. Legal remedies about the above matters are routinely sought even from the Apex Court.

The medical education in India has now become so dear that a student in the private medical college, against payment seat, has to expend about Rs 15 lakhs for becoming a graduate and Rs 25 lakhs for post graduate. It is time MCI/GOI apply realistic norms for opening of new medical colleges and private organisations are encouraged to set up new colleges by attending to their genuine requirements. Similar, emphasis is to be given for setting up more institutions for paramedics.

16. Formation of Rural Health Service and Providing Special Facilities to Doctors & Staff

The doctors always compare their income, living conditions and social life in rural area with those of their counterparts in the urban area. All attempts made by Govt. so far to retain them have not yielded desired results. In view of this it is worthwhile to consider formation of separate Rural Health Service with attractive remuneration to doctors and medical staff. Most of the doctors and medical staff do not stay at the rural health centers. As discussed earlier the absentee rate of doctors and medical staff is very high. In order to motivate them to stay at the centres, it is necessary to provide them decent accommodation along with very good recreational facilities at the place of their posting. There is also another reason for this. Because of emergency nature of duties, it is essential that they should always be available on call. The other difficulty that they face is about the children's education. The education facilities in rural area are not so good. Therefore, they must be assured of the admission of their children in good schools in the nearest town/district place and also hostel accommodation. There is also a need to have proper supervision over the attendance of doctors and medical staff at the health centres, which is not there at present.

17. Setting up National Rural Health Mission

The health and family welfare services have been operating separately even though these are functionally closely related. Also there are number of health programmes operating vertically. This had led to lack of cohesiveness and unsatisfactory results. Govt. of India has therefore launched National Rural Health Mission (NRHM) which will be operationalised from April 2005 with the objective of providing integrated healthcare services to the rural poor.

Besides improving healthcare in rural India, the Mission would focus on decentralisation of management, pool the current "vertically operated health programmes" and utilise the opportunities created by Panchayats for the health scheme.

The Mission's focus will be strengthening primary healthcare through grass root level public health interventions based on community ownership. The mission envisages setting up a health center in every village. The centre will be headed by Asha (Accredited Social Health Activist). Asha will be a representative of the women from the village and will be chosen by the village health committee of the panchayat, the auxiliary nurse midwife (ANM) and the local women's committee. The Mission will be funded through increased budget allocation to Health and Family Welfare as well as pulling the additional tax imposed on cigarettes and other tobacco products.

The Steering Group under this Mission, constituted for policy oversight, is proposed to be chaired by the Prime Minister.

It is not that Govt. has not tried improvements in health services earlier. But these have not met with success. The major shortcomings have been: The Health Delivery System is not accountable to the community it serves. The regular availability of doctors has been a chronic problem. It suffers from lack of funds. As the funds are routed through state governments, there is inordinate delay by the time these reach at the field level; There is red-tape in administration of the schemes due to bureaucratic control. It is hoped that the high power Mission will address all these shortcomings as it runs its course.

11. ADEQUATE ALLOCATION OF FINANCIAL RESOURCES TO HEALTHCARE

The allocation for health and family welfare in 2005-06 (BE) is 10,280 crore. According to United Nation's data, in terms of Govt spending on Public Health, India with spending on health of only 0.9 per cent of GDP ranks 127 out of 175 which is just ahead of Nigeria, Indonesia, Sudan and Myanmar. Thus India is almost at the bottom of the public health spending in the world. In China, with which India is often compared, Govt. spends 1.9 per cent of GDP on health. Table 4.9: Shows GDP per capita and percentage of public health spending in India and China in 2001.

Table 4.9: Health Spending in India and China, 2001

GDP per capita (dollars)	*Public health spending (% of GDP)*
India	0.9
China	1.9

Source: UNDP HDR 2003/2004

The National Health Policy-2002 recognises this fact and one of the indicators incorporated therein is to increase health expenditure by Govt. as a percentage of GDP from 0.9 per cent in 2002 to 2 per cent by 2010. Even this target is too low taking into account enormous requirement in this vital field. It is ironical that as a percentage of national income, on social sectors our country spends lower now in comparison with 1980s.

Presently government accounts for only 20 per cent (0.9 percent of GDP) of the spending on health whereas private spending is 80 per cent (4.2 per cent of GDP. In the cases of hospitalisation, in over 40 per cent of the cases, costs are met by taking loans by the poor or by way of selling assets which are used for livelihood. There are also many reports of closure of already established health sub-centres. The root cause of these happenings is abysmally low allocation to public health services.

In order to improve accessibility of healthcare nearer to villages, there is a need to set up more number of health centres. Also presently rural health centres are not properly equipped. Even most essential medicines are not available in the centres. The main reason for the aforesaid state of affairs is paucity of funds. The allocation for the primary healthcare has therefore to be appreciably increased from the present level. It may be necessary to reallocate the funds between different sectors, giving priority to primary health and basic education.

Further, in view of the importance of Primary Healthcare and Basic Education, levy of a Cess should be considered for raising additional resources required together for health and education instead of levy of cess for education alone. It is hoped that the National Rural Health Mission will address these issues in all seriousness.

12. CONCLUSION

The infant mortality rate in India is very high. Besides, the reports of the deaths of children due to malnutrition appear regularly in the local as well as national press. Although there may be some exaggeration in the reporting and some of the deaths may be related to other causes, the underlying message is that even after 57 years after Independence, there is a lot to be done to achieve the goal of Health for All particularly in the rural area. Large-scale poverty existing in rural area is no doubt one of the main hindrances in improving the health status of the rural population. Lately an idea of covering the poor people under suitable health insurance cover is being toyed. However providing health insurance cover is not likely to solve the main problem of healthcare in the rural area. Against this background, it is necessary to provide the rural people proper primary healthcare which is easily accessible. While there is a steady increase in setting up more and more Primary Health Centres and Sub-Centres, the rate of increase is far too inadequate taking into account the actual requirement. Then there is a problem of getting Doctors who are willing to serve in rural area and retaining them in rural health centres and providing quality healthcare by them. The lack of adequate financial resources and other strategic support to the health centres are also major constraints for equipping the health centres and the expansion and maintenance of the health services. Looking to the enormity of the problems involved, there is a need to adopt certain strategies as mentioned above to achieve the goal of Health for All.

5

Trend Analysis of Budget Allocations to Social and Rural Sectors and How Far these Have Impacted Intended Beneficiaries

1. BUDGET 2004-05: A PRECURSOR

Budget 2004-05, though practically effective only for seven months, is a precursor of the path which will be followed for the next five years. Only those items in the budget 2004-05 are being taken up here, which have direct relevance with our topic under discussion, i.e. Rural Development including Social Sector.

1. Additional Financial Resources for Basic Education and Mid-Day Meal Scheme

Interim Budget 2004-05 provides Rs. 4732 crore for elementary Education. With the objective of providing basic education for all children, it is proposed to levy a 2 percent cess on Central taxes to raise the additional resources to the tune of Rs 4000-5000 crore. Thus, funds totaling about Rs 8732-9732 crores will be used to meet the financial requirements of Sarva Shiksha Abhiyan and Mid-Day Meal Scheme.

2. National Employment Guarantee Act

It is proposed to bring National Employment Guarantee Act with the object of guaranteeing 100 days of employment in a year to one able-bodied person in every poor household. The bill will take into account the experience gained in Maharashtra. Rural Landless Employment

Guarantee Programme (RLEGP) which is merged in Sampooma Grameen Rozgar Yojana (SGRY) provides guaranteed employment to at least one person of every landless household up to 100 days in a year. Since both Employment Guarantee Scheme (EGS) in Maharashtra and RLEGP have met with certain difficulties, it is stated that the Govt. will take care to avoid such pitfalls.

3. Universal Health Insurance Scheme/Group Health Insurance Scheme

Universal Health Insurance Scheme which is in operation has met with little success. The scheme is therefore proposed to be redesigned by lowering the premium and to be made exclusive for persons and families below the poverty line.

It is proposed to introduce a new Group Health Insurance Scheme through public sector non-life insurance companies. The insured will be members of Self Help Groups (SHGs) and other Credit Linked Groups (CLGs) who avail of loans from banks or cooperative institutions. Under the scheme the premium will be Rs. 120 per person and the insurance cover will be Rs. 10.000/-

4. Agricultural Credit

It is intended to double the flow of agricultural credit in three years. The implementation of the policy on agricultural credit announced on 18th June, 2004 is entrusted to public sector and private sector banks, Regional Rural Banks (RRBs) and the Cooperative Banks. During the year 2004-05, farm credit has been enhanced by 30 percent to Rs. 1,05,00 Cr comprising Rs. 57,000 Cr by commercial banks. Rs. 8500 Cr. by regional rural banks and Rs. 39,000 Cr by co-operative banks. The situation of cooperative banking system is termed as grave. In order to find a durable solution, it is proposed to appoint a Task Force to examine the reforms required in the cooperative banking system including the regulatory regime.

5. Debt Relief

Central Govt. announced a package of debt relief and enhanced bank credit on 18th June 2004. The package includes a one-time settlement of loans of all small and marginal fanners in default. This involves a waiver in part, not total write-off. All farmers who are in arrears and thus not eligible for fresh credit, will be provided debt rescheduling and they will also be made eligible for fresh bank credits.

6. Restoring Water Bodies

Through the ages, Indian agriculture has been sustained by natural and man-made water bodies such as-lakes, tanks, ponds and similar structures. It has been estimated that there are a million of such structures and about 500,000 are used for irrigation. Many of them have fallen into disuse. Many have accumulated silt. Many require urgent repairs. It is therefore proposed to launch a massive scheme to repair, renovate and restore all the water bodies that are directly linked to agriculture. To begin with, pilot projects at an estimated cost of Rs. 100 crores, in at least 5 districts (one district in each of the regions) will be taken up.

Once the pilot projects are completed and validated, Govt. will launch the National Water Resources Development Project and complete it over a period of 7 to 10 years. It is suggested to pose the project to multilateral agencies for funding.

7. Water Harvesting Scheme

Water Harvesting Schemes are supported by number of credit institutions. However farmers belonging to the Scheduled Castes and Scheduled Tribes rarely benefit from these schemes. In order to help these farmers, Govt will launch a nationwide water harvesting scheme with an estimated cost of Rs 100 crores. The scheme will cover one lakh irrigation units at an average cost of Rs 20,000 per unit. NABARD will lend money on easy terms and no margin money will be charged from the borrower. Govt. will provide a 50 percent capital subsidy through NABARD.

8. National Horticulture Mission

In a bid to encourage farmers to diversify into area such as horticulture, Govt. proposes to launch National Horticulture Mission with a goal to double the horticulture production from the current level of 150 million tones to 300 million tones by 2011-12. One of the steps is that States will be encouraged to emulate the Anand model and establish a State Level Cooperative Society for promoting horticulture.

9. National Agricultural Insurance Scheme (NAIS)

The Agricultural Insurance Company (AIC) is redesigning the existing National Agricultural Insurance Scheme which is in operation since *Rabi* 1999-2000. Meanwhile, a pilot scheme insuring farm income (in place of crop) has been launched in 19 districts across 12 states during *Rabi* 2003-04. Govt. has decided to extend the scheme to *Kharif 2004* in order to assess its feasibility. AIC is introducing Weather Insurance Scheme on a trial basis in 20 rain gauge stations in this crop season. All three schemes will be evaluated to find the successful one.

10. Traditional Industries

Some of our traditional industries, namely choir, handloom, handicrafts, sericulture, leather, pottery and other cottage industries not only contain great potential for growth and exports, but are integral to our cultural heritage. It is proposed to set up a Fund for the Regeneration of Traditional Industries, with an initial allocation of Rs. 100 crores.

11. Public Distribution System

Fair price shops constitute the backbone of the food security system. There are weaknesses in the system which need to be addressed. In the X Plan document, the Planning Commission had suggested that a system of distributing food stamps should be tested on a pilot basis. Every eligible family will be entitled to collect its monthly quota of food stamps from a designated distribution center, and such stamps could then be used to buy foodgrains from any food shop. It is proposed to introduce a pilot scheme for distributing food stamps, instead of food through fair price shops, in two or three contiguous districts in one state.

12. Subsidies

All the subsidies will be targeted sharply at the poor and truly needy like small and marginal farmers, farm labour and the urban poor.

2. BUDGET 2005-06: A CONTINUITY

There are usual increases in the allocations as part of-every budget. These are not being discussed here. The highlights of the budget 2005-06 with regard to social sector and rural development are:

1. Agricultural Credit

The target of agricultural credit of Rs. 1,05, 000 crore set for 2004-05 is expected to be exceeded by Rs. 3500 crore. Continuing the same path, banks are asked to increase the flow of credit by another 30% in 2005-06. The agency-wise share of credit flow study up to December 2004 shows major share (61%) of commercial banks, followed by co-operative banks (28%) and RRBs (11%). However the share of co-operative banks in 2004-05 is even less than 2003-04 for the same period. This is indicative of the downslide in the performance of co-operative banks.

The recommendations of the task force especially constituted to examine the reforms in the co-operative banks include:

- (i) Special financial assistance to wipe out accumulated losses and strengthen the capital base of co-operative credit institutions;
- (ii) Institutional restructuring to ensure democratisation of institutions; and
- (iii) Changes in the legal framework to empower RBI to enforce prudent financial management.

It is proposed to accept the report in principle.

2. Twelfth Finance Commission

The recommendations of the Twelfth Finance Commission laid before Parliament on 26th February 2004 cover tax devolution, grants to states, debt relief, financing of relief expenditure and related matters. The states stand to gain by these recommendations and they will have some flexibility in providing funds much sought for the social sector and rural development schemes. However, considering the states' finances and their physical responsibility, states deserve more devolution of financial powers than contained in the recommendations.

3. Financing National Rural Health Mission (NRHM)

The focus of NRHM will be strengthening primary healthcare and training of health volunteers etc. To finance the NRHM, apart from increasing the allocation to Health and Family Welfare, it is rightly proposed to tax more of those goods which are health hazards like cigarettes and tobacco products. However these are unlikely to meet the total requirement of the funds for primary healthcare.

4. National Rural Employment Guarantee Scheme (NREGS)

For the Food for Work Programme which has been renamed as National Food for Work Programme, for 2005-06, a provision has been made ofRs. 5400 crores towards cash component up from 4020 crores expenditure in 2004-05 and 50 lakh MT of foodgrains. It is intended to convert this programme into NREGS.

5. Other Special Schemes

All the drinking water schemes have now been brought under the Rajiv Gandhi National Drinking Water Mission. It is intended to cover more habitations and also emphasis will be on tackling water quality. Total Sanitation Campaign (TSC) now in operation in 452 districts will be extended to all the districts in the country.

The diversification of agriculture from foodgrains to fruits, vegetables, flowers, dairy, poultry, fisheries, pulses and oilseeds is being encouraged. The National Horticulture Mission is set to be launched to ensure an end-to-end approach having backward and forward linkages covering research, production, post-harvest management, processing and marketing. It is proposed to introduce a new scheme called Development/Strengthening of Agricultural Marketing Infrastructure, Grading and Standardisation for inducing large investments from the private and co-operative sectors for setting up agricultural markets, marketing infrastructure and support services such as grading, standardization and quality certification. The National Project for the repair, renovation and restoration of water bodies will also be launched in 2005. The pilot project is planned for 16 districts in 9 states and will cover 20,000 ha of additional land.

Under Mission 2007, it is proposed to establish a Knowledge Centre in every village and cover the entire country by using modern information and communication technology (ICT). It is an initiative launched by an alliance comprising 80 organisations. A National Fund for Strategic Agricultural Research is being created to strengthen and modernise the agricultural universities and research institutions.

3. EVALUATION OF BUDGET 2004-05 & 2005-06 AS WELL AS ALL PREVIOUS BUDGETS PRESENTED SINCE INDEPENDENCE

1. Levy of a Cess for Basic Education

There is a need to raise financial resources required for fulfilling the constitutionally mandated basic education programme of Sarva Shiksha Abhiyan and the Mid-Day Meal Scheme. 2 percent cess has been levied in the Budget 2004-05 & 2005-06 to bridge the gap between the funds required and the funds available for the basic education programme. In certain contingencies as mentioned below the application of cess is the only way to raise the resources:

(i) One way of providing additional resources to basic education is to reduce the allocation of some other sectors carrying less priority and transfer the same to basic education. In case it is just not possible to reduce the allocation to any other sector, there is no other alternative but to levy the cess. Thus levying a cess should be the last resort.

(ii) The allocation of cess should be tied up with a specific scheme(s) and detailed account of its utilization should be maintained.

(iii) The allocation of cess to specific scheme(s) should not lead to reduction in the allocation which would have been a rightful share of such scheme(s) in normal course: meaning thereby, these schemes should not become a conduit for indirectly financing programmes/sectors other than the intended ones.

(iv) There should be a separate account for the cess and the cess once collected should not be lapsable if it is not utilised by the end of the year. The unutilised amount should be carried over to the next year which can again be utilized as additional resource and to that extent levy of cess for the next year can be reduced.

(v) Besides Basic Education, Primary Healthcare is another priority area which is suffering for want of funds. Therefore some specific scheme(s) like opening of new Primary Health Centres/Sub-Centres, providing essential equipments at PHC/S-Cs etc should have been clubbed with the basic education schemes and mid-day meal scheme.

In fact levy of cess in the circumstances as mentioned above has been suggested elsewhere in this book. There are numerous instances in the past budgets also when cess or surcharge was levied in special circumstances.

2. Resource Constraints

The social sector (comprising education, health and family welfare, water supply and housing, social welfare and nutrition and rural development schemes) expenditure by Centre amounts to 10 to 11 per cent of total Central spending. The expenditure on social sector by Centre and states taken together amounts to 20 per cent of the total govt. spending. The UNDP recommends a public expenditure ratio, or the percentage of national income that goes into public expenditure of 25 per cent. It suggests a social allocation ratio, or me proportion of public expenditure set aside for social services, of 40 percent. India should in fact go past 40 per cent mark given its poverty level.

Similarly the allocation to agriculture and allied activities amounts to 10 to 11 percent of total budgetary allocation as against the requirement of 20 per cent. The inadequate financial allocation to social and agriculture sectors is the major constraint for achieving the desired progress in the programmes of these sectors.

3. Disbursal of Farm Credit

It is heartening to note mat the target of agricultural credit of Rs. 1,05,000 crore set for 2004-05 is expected to be exceeded by Rs. 3500 crore. Continuing the same path, banks are asked to increase the flow of credit by another 30% in 2005-06. The percentage contribution of co-operative sector in this is however less as compared to the same period last fiscal. The reach of the extension services also needs to be augmented to match with me increased credit flow.

Else it tantamount increased debt burden to farmers without any improvement in his income.

4. Restructuring of Various Schemes

There is a plethora of schemes, some are pretty old and some new ones keep on adding with the introduction of every new budget. The work on some of the old schemes remains unfinished and these are merged with the new ones. There is overlapping of work between different schemes. Rural Development is a complex phenomenon and such a situation complicates the matters further. It is understood that even august body like Planning Commission is finding it difficult to synthesize all these schemes, leave alone difficulty being experienced at the district/block level. It is therefore imperative that all these schemes should be reviewed and restructured suitably.

5. Do Benefits Reach Intended Beneficiaries?

It is crystal clear from the admissions, implicit or explicit, made in the Budget 2004-05 & 2005-06 and admissions made in all previous budgets presented since Independence by the party/parties in power that the benefits from the schemes have not reached to the intended beneficiaries as expected. What are all the reasons and what are the remedies?

Answers to the above problems will be found in this book in the sections on "Strategies to be adopted to achieve the goals" placed at the end of each part.

6

Matrix of the Ingredients of Integrated Approach to Rural Development

So far the goals set in the field of Rural Development could not be achieved because each of the major components that contribute to Rural Development was looked in isolation of the others. Further in this regard there is no common thread binding the Community, Bureaucracy and Political Leadership in framing the policies as well as their implementation. Besides, difficulties have been experienced in achieving the desired goals on account of inadequate allocation of financial resources and not fully and effectively utilising of whatever resources were made available for the purpose. In such a scenario, it is imperative to adopt Integrated Approach towards Rural Development, which alone can succeed in achieving the desired goals.

INGREDIENTS OF INTEGRATED APPROACH

1. Interrelation Between Basic Education, Primary Healthcare & Rural Development
2. Political Commitment
3. Accountability of Bureaucracy
4. Transparency in the Process of Implementation of the Schemes
5. High Absentee Rates of Public Officials & Staff, Teachers, Doctors & Medical Staff
6. Involvement of Rural Community in Planning and Implementation Process
7. Awareness in Rural Masses About their Rights
8. Awareness in Rural Masses About Their Responsibilities
9. Role of Non-Govt. Organisations/Corporate Sector

10. Role of Corporate Sector
11. Member of Parliament Local Area Development Scheme
12. Tackling Important Issue of Population Increase—Influx from Neighbouring Countries
13. Allocation of Adequate Financial Resources to Basic Education, Primary Healthcare and Rural Development
14. Utilisation of Budget allocation
15. Mobilising Addidional Financial Resources Through New Avenues for Basic Education, Primary Healthcare etc.
16. Devolution of Powers from Centre to states
17. Efficient Public Delivery System
18. Restructuring Plethora of Schemes
19. Scientific management of Rural Development Programme
20. Summing up
21. Concluding Remarks.

1. INTERRELATION BETWEEN BASIC EDUCATION, PRIMARY HEALTHCARE & RURAL DEVELOPMENT

1.1. How Education can promote Healthcare and Rural Development

1.1.1. Family Welfare

Family Planning is one of the most important components of the healthcare. However illiteracy prevailing in the rural area is the main hindrance in adopting family planning. Lower rate of illiteracy in girls is further compounding the problem. This is evident from the fact that the birth rate in rural area is much higher than the urban area. Social and cultural background does inhibit the rural people for adopting small family norm. 26 per cent of country's total population is below poverty line and the rural area accounts for three-fourth of the same. There has already been fragmentation of land due to division in families from generation to generation and large size of families means more divisions. This has resulted in large holdings getting reduced to small and marginal holdings, which are not quite as much economically viable. The necessity of family planning is therefore more relevant in rural area. However the specter in rural area is in contrast with the requirement. This situation can change only if the rural people especially girls are educated.

1.1.2. Nutrition

There are quite a number of people in the rural area who do not get regularly two square meals a day. Further, the food that mostly rural people eat lacks nutritive value and this is one of the reasons for their ill health. This particular aspect is all the more relevant in the case of expectant mothers. One of the major reasons for the high infant mortality rate is unawareness about the importance of nutritive food. Education can bring such awareness in the society and this is evident from the fact that there are relatively fewer problems in the urban area where literacy level is higher.

1.1.3. Knowledge about Rights and Responsibilities

Constitution of India provides certain rights and responsibilities to every citizen. Many in the rural India are not aware of these. There are number of programmes/schemes which are in place to help the rural people. These relate to agriculture and other occupations. It is seen that the benefits of these schemes have not reached small and marginal farmers and also weaker sections of the society to the desired extent. The main reason for this is that these underserved people are not fully aware about the benefits as well as procedures for availing the benefits. They miss the benefits unless they are lucky to get assistance from the concerned institutions or non-govt. organisations. Education alone can enable rural population to reap full benefits from the programmes/schemes.

1.1.4. Land Records and Legal Disputes

Those people who are illiterate always face difficulty in getting their land records in order. There are innumerable occasions when the revenue officials at the behest of vested interests tamper their records. These lead to legal disputes and the illiterate people get unnecessarily involved in *Court/Kachheri* cases. These complications can be avoided if the rural population is educated.

1.1.5. Availing Rural Credit

There are various schemes under which rural credit is given to agriculturists. Each scheme has unique features and procedures to get the loan. Unless farmers are educated, they experience lot of difficulties in getting the credit.

1.1.6. Extension Education

Farmers usually adopt the farming practices which they have been following for centuries. Most of them are not aware of the latest research and development that has taken place in farming and farm technology. The State Govt. through agriculture extension staff does provide assistance to the farmers. However, their reach is limited and they cannot cover whole lot of farmers. If rural population is educated they can themselves become aware of the developments through various media.

1.1.7. Employment

The agricultural work is seasonal and therefore agricultural workforce is underemployed. If they are educated they can either start some venture in rural area itself or else move to urban area where they can find suitable employment.

1.1.8. Empowerment of the Rural Population

Since rural population constitutes almost three-fourth of total population of the country, India's progress is closely linked to human resource development in rural area. This kind of development is just not possible with the present rate of illiteracy. In this context, increasing the out reach of literacy to cover even the remotest part of the country is the top priority.

1.2 How Healthcare can promote Education and Rural Development

1.2.1. Mid-day Meal Scheme in Schools

Save some persons with large land holdings or side business, there is utter poverty in rural area. Below Poverty Line families in rural population find difficulty even for getting the food and their children are undernourished. In this context Mid-day Meal Scheme in the schools has become popular as it provides needed nourishment to the children as well as it acts as a motivator for enrolment of the children in the schools.

1.2.2. Improvement in Attendance of Children in Schools

The enrolment and regular attendance of the children in rural area, among other factors, is dependent on the well being of the children. There are also instances when children have to stay away from their classes when they have to attend to their family members. The factor of ill health is more predominant in rural area as compared to urban area as proper healthcare facilities are not available at hand in rural area.

1.2.3. Incidence of Diseases and Epidemics Affecting Rural Population

The incidence of diseases and epidemics are predominant in rural area as proper sanitation facilities, safe drinking water, nutritious food and proper healthcare facilities are not available. This directly affects health of the rural population and importantly their productivity. Another factor is that there are no separate cattle sheds and the people reside in close proximity with the cattle. This also is a cause for many diseases.

1.3. How Poor Rural Development can Inhibit Education and Healthcare

Here Rural Development has a meaning of Rural Economy.

1.3.1. Poverty Inhibits Enrolment and Attendance of Children in Schools

26 per cent of country's total population is below poverty line and the rural area mostly accounts for majority of BPL families. The poor families are preoccupied with such activities that will give them food. All other activities including education are lower in their priority. This has been a main reason for not enrolling their children in schools. Against this background the introduction of Mid-day Meal scheme has evoked excellent response in poor families and there is an improvement in the enrolment as well as children continuing education.

1.3.2. Expenditure on Healthcare

In view of the widespread poverty prevailing in the rural area, the quantum of expenditure on healthcare in rural family forms a very small part of the total expenditure. Rural people seek healthcare and spend money on it only in dire circumstances. On the other hand in urban area, people spend not only when they fall ill, but many of them spend also on fitness centers and gyms even when they do not suffer from any illness. If there is all round development in rural area and its economy improves, rural families will be in a better position to seek proper healthcare and spend necessary money on it as and when they need it.

2. Political Commitment

2.1. Inter Sectoral Priority

There are various sectors at Central as well as State level claiming priority when it comes to policy formulation as well as allocation of funds. It is here that the political commitment to Basic Education, Primary Healthcare & Rural Development becomes necessary. These areas suffer from many angles. There are no strong organisations/ associations representing people working in these areas. If they are there, they are disparate. Another aspect is that these areas cannot be looked from the narrow point of view of generating incomes and profit although others make profit out of the produce supplied by the rural people. On the other hand from the view point of social justice and balanced development, the priority of basic education, primary healthcare and rural development overrides other priorities. In the light of this it is very necessary to have political commitment for improving the level of Basic Education, Primary Healthcare & Rural Development even if it means transferring the resources to these from lesser priority areas.

2.2. Empowering Panchayat Raj Institutions

The schemes are planned, targets are set and allocations are made to achieve the targets. However it is seen that the targets in the area of Basic Education, Primary Healthcare & Rural Development are not met. This happens mainly because the end point where the execution of the scheme is being done, the persons who are executing the scheme are not either adequately empowered to take decisions or provided with timely funds required to execute the scheme. It is therefore necessary to empower, in real sense, the rural organisational structure i.e. Gram Panchayat/Gram Sabha, Panchayat Samiti or Zilla Parishad, as the case may be, to achieve the desired results.

2.3. Reducing Political Interference

Political interventions in the situations where things are not going in right direction are not only fully justified but should be welcomed. Rather the persons responsible for the underperformance should be taken to task. However if the process turns into interference in day-to-day functioning of the civil service, there will be a dilution of the effectiveness of the latter to deliver the results. The matters of transfers and postings of service officials and staff are often used as a tool by the political class to serve its interests. However for this, political class alone cannot be blamed because service officials and staff themselves also often approach the former to seek favours for postings to the places that offer higher status, perks and money.

In this context, in Karnataka, they have limited the transfers, posted transfer data on website and formed management committees to examine the transfer requests. In Maharashtra State also guidelines have been formulated for effecting transfers. Other states too need to formulate guidelines for transfers.

3. Accountability of Bureaucracy

3.1. Implementation of Policies/Programmes

All policies/programmes as approved by the Parliament/ Legislature are to be implemented through organisation structure called bureaucracy/civil service. The literary meaning of bureaucracy is a set of officials. Although in real terms Gram Panchayat, Panchayat Samiti or Zilla Parishad execute the programmes/schemes, the required inputs like staff, funds, technical support etc are to be processed and provided by the bureaucracy. If funds are not received in time it is just not possible to implement the scheme. As such the success of the scheme is dependent on the sincerity of the bureaucracy in providing the required inputs. It is generally observed that schemes suffer on account of typical bureaucratic approach of the concerned officials when it comes to providing necessary inputs to field organisations. The situation can improve if bureaucracy is also made equally accountable for the implementation of the schemes along with the field organisations.

3.2. Reforming Bureaucracy

During pre Independence era, the main preoccupation of the civil service was to maintain public order. But after independence the needs of governance have changed. Besides public order, development aspect has assumed great importance which was not there earlier. During post Independence period, the nature of polity has considerably changed and called for a similar transformation in the civil service. However the basic character of the civil service which has been inherited from the colonial era has not undergone the same change. The Babus have particularly remained indifferent to the change process. The democratisation of the political process and the social churning it has caused, requires to be correspondingly supported by the civil service. In essence it requires improvement in governance and efficiency. This calls for a change in the present style of functioning of the civil service and adopting participative approach of governance. It will make a lot of difference if the civil service treats the people as their customer. This change for good is already visible in the working of the public sector banks, Indian Airlines etc. However it is also true that much depends upon the attitude of the person working in the civil service. Many of them are very popular in the public because of their positive response to the genuine needs of the people.

Although the overall number in staff strength is all right, the staff in the administrative departments can be reduced and assigned to field units. Also number of administrative departments can be restructured by abolishing unproductive ones so as to improve coordination, reduce time delays and result in better public delivery.

To ensure proper performance, there are occasions when it becomes necessary to take disciplinary action against some officials. The present procedure for taking disciplinary action is quite cumbersome. These procedures need to be simplified. Similarly the honest and competent officials should be rewarded.

3.3. Sensitivity Training

Rural people have a general feeling that the officials and staff of the State Govt/Agencies are indifferent to their problems and they do not get proper hearing of their grievances. To improve this perception, there is a need to sensitise those officials and staff with whom the rural people have to generally interact through suitable training. It should include both components i.e. pre-induction as well as post induction refresher training.

There will also be better appreciation of the difficulties being faced by the field organisations, if those in the civil service are exposed to field duties by way of posting them in the rural area say for 2-3 years after every 5-7 years of posting in town/city.

3.4. Right to Information

Often the rural person concerned does not know the exact status of his case which is pending in the office. There is no way to ensure passing of the necessary information to the concerned persons without enacting the law for Right to Information. Such enactment would ensure accountability of the person dealing the matter as well as take away lethargy presently witnessed in dealing with the cases. Maharashtra state has already enacted such a law.

Information obtained by the Times of India (21.03.2005) newspaper from the Maharashtra Mantralaya through the Right to Information Act – reveals that in December 2004, files pending numbered 1.89 lakhs and the rate of disposal and the average cost per file is:

Dept	*Salary for Dec. 04 in Rs. lakhs*	*Files cleared Nos.*	*Avg. cost per file in Rs.*
General Adm.	177	524	33,806
Home	58.36	2511	2324
PWD	64.72	733	8830
Finance	57.78	2677	2158
Irrigation	65.43	1078	6070
Revn. & Forst.	63.29	1204	5257

Source: TOI dt. 21.03.2005

If the state of affairs in relatively progressive state like Maharashtra is so bad, it could be pathetic in many other states. A time bound clearance of files, elimination of needless paper work and accountability should be made mandatory. Also it should be mandatory for department heads to conduct periodically a stock-taking of all pending files and ensure speedy disposal.

4. Transparency in the Process of Implementation of the Schemes

Our country stands globally very tall in the scale of corruption. Berlin based anti-graft group Transparency International measures corruption by what is called corruption perception index (CPI). The index defines corruption as the abuse of public office for private gain, and measures the degree to which corruption is perceived to exist among a country's public officials & politicians. As per Transparency International Report 2004, in a total of 146 nations, India is ranked at a lowly 90. It is interesting to note that generally only "haves" indulge in malpractices that lead to exploitation of "have nots" i.e. people like rural folk, which proves the point that the corruption breeds out of greed and not need. In countries like Europe or USA, in their democratic set up although corruption is prevalent in certain quarters, interestingly average man is not so much affected. In our country there are situations when it is difficult to get things done without advancing money even when they fit into the rule. In this respect the rural people are the worst sufferers as most of them are illiterate and can be exploited easily. It is very difficult for them to get anything done without advancing the money, save exceptions.

As a matter of fact, with the periodic awards of the central pay commissions and subsequently adopting the awards on similar lines by the state governments, the remuneration of public servants has gone up considerably. It is more than what is generally paid by the private sector for the same amount of work. Therefore the argument that the person is indulging in the act of corruption because he is poorly paid does not hold the water. The concerned official does not even admit this as corruption, instead he terms it respectfully as *Suvidha Shulk.*

India's high ranking in corruption means our present democratic institutional framework has failed to provide good governance. There is a general perception that lower rungs of administration get cue from their higher ups and therefore the process of cleansing should start from top down i.e. political class as well as bureaucracy. In this context, there is a need to strengthen internal audit procedures and adopt comprehensive anticorruption strategy.

There should be transparency in the working of the bureaucracy. It becomes quite helpful if salient features of the procedure, for any public work to be done, are highlighted by way of displaying them prominently on the office notice board. Further if there is accountability of the individual's performance, the incidence of the malpractices will be minimised.

There is a lot more scope for improvement in the working of the field organisations, which are supposed to implement the schemes. There are numerous examples of awarding the development work by the Panchayat Raj Institutions (PRI) to unqualified organisations who ultimately deliver sub-standard work. The only remedy to avoid such happenings is to incorporate transparency in the working of the PRIs. This is possible if rural masses in general are aware about various programmes/schemes under implementation and they have a right to information. Maharashtra state has already made such a provision but it is yet to be tested for its actual implementation.

In Nagpur (Maharashtra), District Administration recently launched a project called *Adhar*. Under this project, the beneficiaries of the welfare schemes are issued smart cards similar to an ATM with an advanced feature of a thumb impression of the card holder, for disbursal of cash. The project is aimed at eliminating many bogus beneficiaries as well as middlemen who were getting paid hefty sums of money. Similar innovations need to be initiated in rural *area* to avoid exploitation of rural people. Corruption is endemic and it has to be tackled on all fronts. Even other professional people are equally to be blamed. What is needed is to adopt proper work culture. Gandhiji has very well put it that everybody should act as a trustee of the business under his charge. If this principle is followed in all facets of life, life of a common man will become quite comfortable.

However it would be unfair to paint all people's representatives, bureaucrats and other professionals with one brush. In fact it is because of the positive contribution of good people's representatives, bureaucrats and other professionals that some development is visible and the common man is able to sustain the difficulties.

5. High Absentee Rates of Public Officials & Staff, Teachers, Doctors & Medical Staff

Non-Attendance/Non-Punctuality of officials and staff of the State Govt/Zilla Parishd is a major problem being faced by the rural population. It is estimated that the teacher absentee rate is to the extent of 15-26 per cent and the rates for doctors/paramedical staff twice that. Here when we talk about the absence of officials/staff, they relate to all departments like civil administration, electricity Board, schools, PHCs/Sub-Centres, etc. They include among others- teachers, patwaris, doctors/para-medical staff, linemen etc. Most of these staff members reside at the places other than the place of their postings and generally punctuality is not adhered to. This affects their performance on the duty. Also, since the absentees are on pay rolls of the State, the practice tantamount to wastage of public money. To this if the rate of non-filling of vacancies is added, the rate of delivery of public service is very poor.

As regards grievances relating to outstations, the villagers cannot afford to be away from their work places often and they have no means to leave the village. As such they are the worst sufferers. This menace can only be tackled by regular visits/inspections of the work places by the higher officers and taking strict action against the defaulters. In this context, Gram Panchayat can also be empowered to ensure punctuality.

Taking into account the emergency nature of duty, there is a need to provide proper residential facility for the medical staff at the PHCs/ Sub-Centres. This is likely to motivate them to be regular in their attendance and punctuality. Also provision of dormitory-type accommodation for the teachers as a part of the school premises could also be considered.

6. Participation of Rural Community in planning and implementation process

The main reason for failure to achieve the targets set in the field of basic education, primary healthcare and rural development is preparing plans without taking into account ground realities. Rural community knows what is good for them. They should be involved at the time of detailed planning for the schemes. In planning, old top down approach has to be replaced by the bottom up approach. The former approach has in the past failed tc deliver the results whereas the latter wherever it has been tried, results have been encouraging.

7. Awareness in Rural Masses about their Rights

There is general lack of awareness in rural masses about their rights and responsibilities. There are many programmes/schemes which are in various stages of implementation by the Govt. but rural masses are not fully aware about these schemes as a result of which they are not able to reap full benefits from these schemes. The service staff i.e. gramsevak, extension worker, health guide etc. can educate the villagers about such matters. NGOs can also play important role in this respect.

8. Awareness in Rural Masses about their Responsibilities Towards Society

It has to be realized that along with the rights, rural masses have also certain responsibilities towards the society. This can be exemplified by the fact that many a times schemes are not put to proper use by some of the people for whom the schemes are made. There are many examples like the loan taken for digging the well or purchasing the cattle are used for some other purposes which include even celebrating family ceremonies. Also some people do not make earnest efforts to return the loan in time.

9. Role of Non-Govt Organisations

In a vast country like India, the task of achieving the goal of education for all, health for all and all round development of rural area is so gigantic that Govt. alone cannot handle it. It requires multilevel and multidimentional approach and in this context the role of NGOs assumes great importance. There is in built red tape in the working of the Govt. organisations and it helps if the work is assigned to societies and Govt. acts only as fund provider and facilitator. There are many schools now being run by societies in rural area on this principle. There are however problems even with the societies. There are many NGOs whose working is unsatisfactory. It is necessary to constantly monitor the working of the NGOs and de-registering those whose working continues to be unsatisfactory. In this context there is a need to benchmark the performance level of the NGOs for monitoring purpose. The NGOs can be graded as per their performance and the funds should be provided accordingly.

However many a times the departmental officials co-ordinating the work of NGOs are hand is gloves with the NGOs for pecuniary benefits and the performance of NGOs is not critically assessed.

There are 12 lakh NGOs in the country, out of which only 50 per cent are registered. It is necessary to streamline the working of NGOs. The Centre is already considering setting up a regulatory body for NGOs. However the better way will be to form a database for NGOs and frame the guidelines about how best the services of NGOs can be utilised. As mentioned, at present NGOs are forced to keep *Babus* in the Govt. in good humour. A latest budget (2005-06) proposal allowing NGOs engaged in micro-finance to access the overseas debt market could make NGOs to look beyond Govt. grants and shape their working independently.

10. Role of Corporate Sector

Today every nation in the whole World is concentrating on solving the critical issues being faced by their people. The Corporate Sector has a special responsibility to contribute to this larger mission. The Indian corporate sector, while generating economic surplus, needs to focus on replenishing natural capital and on eliminating poverty through the creation of sustainable employment opportunities. Generally corporate social and environmental initiatives tend to be philanthropic. But many corporates now have started getting actively involved in activities related to rural development. To name a few, NTPC, Tatas, ITC etc. have already formulated certain policies towards achieving these objectives. Since it is impossible for the Govt. alone to achieve the all round development of vast rural area, there is urgent need for all corporates to join this noble task.

11. Member of Parliament Local Area Development Scheme (MPLADS)

The normal process of planning is governed by a set of laid down guidelines with regard to cost norms, resource allocation and prioritization of works to be undertaken. Consequently, it is not always possible to take care of some of the pressing felt needs of the people in all circumstances. It was with a view of bridging this gap and providing certain degree of flexibility in the planning process for meeting the pressing needs of the people that the Member of Parliament Local Area Development Scheme was started in December 1993. The scheme is governed by a set of guidelines.

Some of the salient features of the scheme are: (i) The work under the scheme shall be developmental in nature, for creation of durable assets; (ii) Each work suggested under the scheme should not normally cost more than Rs 25 lakhs; (iii) Public utility works of registered societies subject to certain conditions are also permissible; (iv) Certain works such as office buildings, repairs and maintenance works, creation of assets for individual benefits etc are not permitted under the scheme.

This amount is Rs. 2 crores per year for each MP. However the average utilization of the amount released (10,498 crore) as on 31st March 2003 is only 78.8 per cent. Audit reports have pointed out that there is no proper accounting system in place for this scheme. The nodal Ministry at the Centre sends money directly to the district collector who after getting the sanction from the MP gives the money to the implementation agency. After that there is no proper monitoring of the work in progress. Further it is observed that out of the works executed, the money under this scheme has predominantly been spent on roads, bridges and construction works, while schemes pertaining to health, education and sanitation received relatively low priority. Local Level Committees should be formed to suggest the priorities of works, monitor the progress of the implementation of the schemes and play a watchdog role so that the schemes envisaged are not stymied by leakages.

There are different views about the continuation of the scheme. The reports of the National Commission to Review the Working of the Constitution (NCRWC) and Institute of Social Sciences (ISS) have opined to abolish the scheme citing different reasons such as it goes against the demarcation of responsibilities between the legislature and the executive and that the scheme has not achieved what it was expected to. In any case there is a definite need to frame new guidelines for the scheme with emphasis on greater transparency, better utilisation of funds and expeditious execution of projects.

12. Tackling Important Issue of Population Increase—Influx from Neighbouring Countries

Any Census or Survey data has to be handled and interpreted carefully and more so population census as in the case of India it encompasses more than one billion people spread across length and breadth of the country and all this is done with limited resources of enumeration manpower. We have already discussed in earlier chapters about the strategy to be adopted for controlling the internal population increase. Further, the influx of people crossing over unofficially from neighbouring countries cannot be ruled out. There are already strong reports from north-eastern states about the change in the whole demographic picture in rural country side. Since India is already grappling somehow with solving the problems of rural development programmes for its existing population, it can ill afford influx from other countries. In view of long land border, fencing can reduce the influx to some extent but by itself it cannot be the holistic or durable solution. Such cross over is taking place in hordes even from Mexico to USA on regular basis even though the latter adopts stringent measures against such cross over. We should work on the real solution which lies in helping the neighbouring countries to the extent possible towards alleviation of abject poverty in those countries. Such help need not be thought of from financial angle alone. It can be technical, offering management consultancy or sharing our experiences of development programmes with them. All this does not mean we should slacken our effort on fencing or guarding our border, which is still one of the most important steps to curb the influx of people.

13. Allocation of Adequate Financial Resources to Basic Education, Primary Healthcare And Rural Development

In order to increase the out reach of primary education, new schools are to be opened. In the existing schools, necessary facilities are to be provided. For improving the accessibility of healthcare nearer to villages, more number of primary heath centers/sub-centres are to be set up. For improving living conditions in rural area and rural economy, necessary infrastructure like roads, safe drinking water, shelters, marketing & credit facilities etc are to be provided. It is seen that most of the programmes/schemes are languishing on account of paucity of funds. The allocation to Basic Education, Primary Healthcare and Rural Development is therefore required to be suitably enhanced. There has been increase in the allocation of funds every year in absolute terms. However the allocation as a percentage of GDP on social sector has been dwindling from year to year. Even the increases that are seen in absolute terms are on account of steep hikes in salaries of the Govt. employees as a result of Central Pay Commissions' recommendations and their cascading effect on states and state dependent agencies like private schools of whose teachers' salaries are paid by the Govt.

Since staff salaries and administrative expenditure alone consume almost 80 per cent of the allocation being made to education or health sector, there is a strong need to contain periodic steep hikes in the salaries if adequate money has be secured for the delivery of service at the end point. Otherwise all these schemes will serve as instruments for giving employment to teachers and medical staff etc. and not being able to achieve the goals of education for all, health for all and rural development due to non availability of funds for requirements other than paying staff salaries.

14. Utilisation of Budget Allocation

For Centrally sponsored programmes, the Centre makes the allocations under various heads but once they reach the state they loose their identity in the larger pool that also comprises Central transfers under the Gadgil formula and the funds of the state. The problem has worsened with the growing administrative expenditure and vested interests exercising control over the funds. With the implementation of every new Pay Commission's recommendations, the administrative expenditure is hiked. Of the Rs. 5,000 crores set aside for the centrally sponsored employment generation programmes, utilisation is not more than 50 per cent of the budgetary allocation. Of the 50 per cent supposedly to be utilised, only 40 per cent is actually deployed towards the actual project. This is reportedly revealed in the 'accounts' put together by the Reserve Bank of India two years after the budgetary allocations are made. Thus finally only 20 per cent of the allocation made by the Centre reaches the actual beneficiaries.

The rest is a loss to the schemes due to red tape in bureaucracy resulting in delays in providing various inputs and release of funds, infructuous administrative expenditure and leakages of funds at various levels in the chain of processes involved in implementation of the schemes.

As regards utilisation of allocation let us take the example of District Primary Education Programme (DPEP). The cumulative allocation for DPEP during the FYs 2001-02, 2002-03 and 2003-04 was Rs 5847.2 crore and utilisation was only Rs 3275.1 crore which is just 56 per cent of the allocation. The utilisation in 2003-04 was however 64.2 per cent of the allocation. For Mid-Day Meals, during the same 3 year period, out of the allocation of 84.59 lakh MT of foodgrains, 63.09 MT was lifted, which amounts to utilisation of 74.6 per cent of the allocation.

It will be unfair to put the blame on state level alone. There are delays in transfer of funds at every stage whether it is Centre or State and within the State. This is happening purely on account of bureaucratic approach of the govt. departments involved. Since the implementation of the programmes is now being devolved on Panchayat Raj Institutions (PRIs), there is a proposal to transfer the funds of

Centrally sponsored programmes directly from Centre to PRIs but via State concerned. However in our federal structure, it will be difficult to get such a proposal accepted. Besides, instances of nonutilisation or misutilization of funds sometimes take place even at Panchayat level.

With advanced Information Technology available in India it should be possible to regularly monitor the physical progress of the schemes as well as utilisation of budget. Once these inputs are available, it should become easier to take remedial measures if there are slippages.

15. Mobilising Additional Financial Resources Through New Avenues for Basic Education, Primary Healthcare etc.

15.1. Levy of a Cess

Basic Education and Primary Healthcare are core responsibilities of the Govt. The allocation of funds between different sectors may be done with reference to priority and if necessary reallocation of funds may have to be done. Even after doing this if there is a shortfall, it is worthwhile considering new avenues to garner the financial resources. One of these could be through levy of a Cess as a percentage of the Central taxes. This is similar to a surcharge levied after Gujarat Earthquake.

15.2. Foreign Investment in Basic Education, Primary Healthcare and Minor Irrigation

India having adopted "Reform Regime" there should be no difficulty in accepting foreign investment in the field of elementary education and primary healthcare. However for this, it is necessary to formulate suitable policy so as to attract foreign investment.

Similarly out of five lakh water bodies, only half are usable and the rest need repair, renovation and restoration. Since the investment required for this is quite high, multilateral funding especially by the World Bank can be pursued.

15.3. Adopting Subsidy Reforms

Prepared in collaboration with National Institute of Public Finance and Policy, GOI report placed in Parliament on 23. 12. 2004 states that the total subsidies at Rs. 1,15,824 Crore in 2003-04 accounted for 4.18 per cent of GDP of which non-merit subsidies at Rs. 67,250 crore accounted for 58 per cent of the total bill. Merit Subsidies comprise expenditure on primary education, flood control, road and bridge construction, sewerage and sanitation services, whereas subsidies on power, irrigation, transport, higher education, agriculture and industry are considered non-merit. The report states:

"The subsidy reforms should aim at eliminating non-merit subsidies and introducing uniform price policy with a system of food coupons for Below Poverty Line (BPL) families as the present system of dual pricing under PDS encouraged leakages.

Stressing that fertilizer subsidy in its present form be done away with, the urea should be decanalised, flat-rate subsidy system be introduced and a mechanism to increase farm-gate price of urea at regular intervals be considered. The domestic LPG and PDS kerosene subsidies seems to be ineffective in serving the desired objective and hence domestic LPG subsidy may be gradually reduced and a cautious approach pursued. Only state-owned oil companies have been permitted to market subsidized domestic LPG and PDS kerosene, which has stifled competition by restricting the entry of private retailers. A market environment encouraging fair and healthy competition is the most effective way to expand the supply and availability of competitively priced kerosene and LPG."

There appears to be a need to redefine the division between merit and non-merit subsidies such as agriculture by itself is not sustainable and it should be included in the list of merit subsidies. The above recommendations thus can be considered by the Govt. subject to the maintenance of a social safety net for the rural population. Rather the cut in non-merit subsidy can be channelised to programmes of basic education, primary healthcare and rural development, which often suffer for want of funds.

15.4. Charging Ground Water Use

A large quantity of groundwater is accessed by soft drink and bottled water manufacturers. These manufacturers including MNCs are earning huge profits. However there is no charge for accessing the groundwater. The companies should be asked to pay for the quantity they use and this amount can be earmarked for allocation to Primary Healthcare and Basic Education.

15.5 Use of Forex Reserves for Infrastructure Development

India is faced with the problem of meeting the huge requirement of funds for infrastructure development. It is estimated that India needs $ 150 billion in infrastructure over the next 10 years. It should be possible to draw $ 15 billion from our foreign exchange reserves to fund infrastructure development. The $ 15 billion can be used as seed money in the shape of equity for leveraging 10 times this amount in the form of debt and equity. As an example, Chinese Govt. recently used $ 45 billion of its foreign exchange reserves like this for infrastructure development.

16. Devolution of Power from Centre to States

Just as implementation of schemes is being delegated to Panchayat Raj Institutions for better results, there is a need for Central Government to devolve greater financial powers to states and PRIs. In a democratic federal set up, this is inevitable. No doubt, there is always a danger lurking in the eyes, that such powers may be misused at the state level by those who wield these and there are numerous examples of these around. In PIL case regarding transfer of mid-day meal scheme from Centre to States, the Apex Court has shown its disfavour to such a proposal in the schools from Center to States.

Notwithstanding situation as mentioned above, there is no way to expect all round progress in a vast country like India especially in rural area without delegation of adequate powers to the states and institutions that are nearer to the people. Such increased devolution has in fact been suggested by Sarkaria Commission as well as various Finance Commissions appointed by the Central Govt. from time to time. At best there can be suitable guidelines and superintendence for use of these powers as and when these are delegated.

Meanwhile GOI on 1st February 2005 has accepted the recommendations of the 12th Finance Commission headed by noted economist C. Rangarajan in toto as an "award" under which the states would get a one per cent higher share at 30.5 per cent in central taxes and duties, pay a lower interest of 7.5 per cent, down from 9 per cent now on all central loans and can have their total debts written off in the next five years by implementing fiscal parameters.

The total impact on account of implementation of TFC recommendations on the Central budget for 2005-06 will be Rs. 26,000 crore or an addition of three-quarters of a percentage point as a proportion of GDP. This will have an impact on Central Government's capacity to abide by the Fiscal Responsibility and Budget Management Act (FRBM). However the expenditure is justified if the states pass similar FRBM and abide by it.

17. Efficient Public Delivery System

Since Independence, the Govt. started number of good schemes/ programmes for the welfare of the rural community. Lot of money has already been spent on these schemes. However this has not impacted significantly the pace of development in the rural area. The major reason for this is that efficient delivery system has not been in place with the result that the benefits have not reached the intended beneficiaries. For example, it is relatively difficult for a small and marginal farmer to get the loan from the bank, although he needs it most. He therefore gets it from the private lender at higher interest charges which increases his debt burden.

Farmers face difficulties in getting essential input i.e. quality seeds in time. There are many substandard seeds floating around in the market and it is difficult to identify these. Also many a times quality seeds arrive quite late in the market. This kind of uncertainty compels farmers to buy substandard seeds as they cannot afford to miss the date of sowing, which is solely dictated by the rains. Similar difficulties have been experienced in the availability of fertilizer i.e. urea in time. At times when farmers get supplies late, they use urea in more than the required proportion assuming that using extra quantity will make up for the delay.

Govt. or even Panchayats should free themselves from actual execution of the schemes, unless they are small ones. It can best be left to competent Non-Govt. Organisations or private organisations. Govt. or Panchayats should function more as regulatory bodies or facilitators and ensure that the work goes on smoothly.

If any job is to be got done by the villager, it is not done easily. Many a times he has first to advance money, not officially. It is generally difficult to get even Kisan Credit Card by the poor farmer without advancing money. It is therefore essential that whenever any scheme is to be started, the necessary delivery system should be evolved and put in place right at the time of starting the scheme. In this respect, computerisation of the various processes from planning to implementation can be very helpful. If efficiency and transparency are built in the system, it is bound to succeed. Allowing right to information will also help greatly. There will be no doubt stiff resistance for effecting the above changes from the vested interests, but it has got to be surmounted.

It is generally experienced that the high absentee rates of State

Govt/Zilla Parishad employees/school teachers/doctors and medical staff/revenue staff etc. from their duty places affect the public delivery system badly. Public woos are further compounded by unfilled vacancies in the schools/health centres.

Improvement in delivery system requires reforms specifically in the bureaucracy as it has a real control over the support system that is crucial for achievements at the field level.

18. Restructuring of Plethora of Schemes

There are more than 200 Centrally- sponsored schemes. Many of the schemes, though well intentioned, remain only on paper or incomplete. Some of the schemes undergo not only change of name but also basic character. Many a times new schemes are introduced and old ones are merged mid way in the new ones, even though works of the old ones are incomplete. Some of the schemes are running through many ministries and these schemes in course of time are treated as orphans i.e. without *Mai-Bap.* There are also schemes overlapping each other. It appears even Planning Commission is finding it difficult to keep track of all the schemes.

Central Govt. spends Rs. 39,000 crore annually on 200 Centrally Sponsored Schemes (CSS). Earlier as an exercise of restructuring, 80 schemes with an expenditure of Rs. 5,000 crores were identified for either weeding out or merging with other schemes or transferring to the states as it was thought that they have outlived their utility. But there is now a consensus in principle for discontinuation of at least 7 schemes, keeping other schemes alive.

It is high time to take a close look at the plethora of schemes, their present status and usefulness, and restructuring them wherever necessary. Further there is a need to promote local area planning. Thus restructuring of these schemes involves many challenges like: rationalising loan-grant ratio, unbundling or consolidating the schemes, willingness of states to accept the changes, willingness of bureaucracy to accept a diminished role if Central schemes are unbundled or consolidated and the capacity of local bodies to do the job. Notwithstanding the difficulties involved, restructuring of the schemes deserves great urgency. In fact National Development Council Sub-committee has been assigned the task of reviewing the schemes. In this context the work has already started and the first one was expected to be given effect to was mid-day meal scheme, a project managed by the Centre. However citing its criticality and importance, the Apex court has presently stay put the proposal.

19. Scientific Management of Rural Development Programmes

Many of our rural development programmes have failed because they were not managed properly. If they have to succeed, they need to be managed on scientific lines. There are several definitions of management. However the definition which comes closest to our related subject matter i.e. rural development is given by Mary Cushing Niles as, "Good management, or scientific management, achieves a social objective with the best use of human and material energy and time, and with satisfaction for the participants and the public".

The analysis of various characteristics of management reveals that it is a process of planning, organising, coordination and control. For the best results, these functions are to be performed on scientific lines.

Planning

According to Management expert — Hart, Planning is determining in advance of a line of action by which certain results are to be achieved. At the national level, there is a need to specially focus on planning for rural development progrmmes as it is impossible to achieve rural development in the natural course. For scientific planning, there is a need to clearly set the objectives like what is the mission and purpose of the programme. The planning is deciding in advance what, where, and how the work is to be done. The planning process involves breaking down the objectives into goals for each unit, laying down policies, programmes, procedures, budgets, strategies and schedules, all directed towards attaining desired results.

The plans can be long term, short term or for weeks/days. The long term plans are formulated by the top level executives, short term plans by the middle level executives and weekly/daily plans are formulated by the lower level executives. The short term plans are required to be correlated to long term plans and similarly weekly/daily plans are to be correlated to short term plans.

The planning process involves establishing clear objectives, identifying real problems in achieving the objectives, collection of facts and data, identifying alternative courses of action, evaluation of alternative courses of action, choice of a course of action, and formulating derivative plans for different departments.

The planning function being the starting point in the whole process of achieving the goal, it has to be realistic and for this purpose it should be based upon discussion at various levels. A programme may consist of number of schemes or projects. In the context of rural development scheme, it is necessary to first make survey of the actual needs of the villagers and prepare a draft plan for the scheme. Then this draft plan should be discussed with the leaders/peers of the village so that the plan takes into account the problems likely to be encountered while implementing the scheme. If such a procedure is followed, there is a less chance of the failure of the scheme.

Each scheme should specify the following basic information:

- Title of the rural development scheme
- Objective/purpose of the scheme
- Requisite details of the socio-economic survey conducted
- Area of coverage/target group to be covered
- Schedule for the implementation
- Physical and financial targets
- Cost-benefit ratio/social benefits expected
- Funding required and how it is proposed to be made available
- Organising/staffing for implementation and identifying core group/person
- Arrangements for providing inputs-critical and non-critical
- Linkages with other schemes
- Establishing mechanism for—Monitoring and control
- Conflict resolution mechanism
- Independent evaluation (with criteria of evaluation) depending on the size and importance of the scheme

Presently there is a Planning Cell at the district/block level, Planning Board at the state level and Planning Commission at the central level. In spite of this organisation set up, the planning lacks requisite qualitative and detailing aspects as is being done in industry sector. Planning for rural development programmes is macro-oriented, flowing from set national goals and priorities. However they very much lack micro-planning required for the implementation at the field level.

Organising/Staffing

Organising means providing organisation structure and staffing appropriately. The organisation is thus composed of people *who* run and manage them. There are numerous definitions of organisation. However, the definition given by Johnson, Kast and Rosenzwerg fits well with our subject matter. According to it, 'the organisation is the assemblage of people, materials, machines and other resources geared to task accomplishments through a series of interaction and integrated into a social system"

The scope of designing the organisation should not be limited to spell out division of the tasks, chain of command or the line of authority and logical relationship of various functions. It should also ensure providing environment for each person in the organisation to effectively participate in the overall organisational effort. Presently what we witness in Govt. institutions is the bureaucratic form of organisation which in layman's term refers to the negative characteristics of large organisations, such as excessive red-tape, procedural delays and general frustration. It emphasises too much on formal procedures, rules and regulations.

Thus design of organisation has to take care of:

- It should facilitate effective coordination of macro as well as micro-planning i.e. vertical integration.
- It should provide horizontal integration so as to obtain the best cooperation from other related disciplines.
- It should have staff with professional competence necessary for implementation of the scheme.
- It should adopt participative style of functioning both within as well as when the function is related with outside organisations.
- Its staff should be sensitive to the needs and problems of the rural poor especially illiterate people.

In the context of rural development, organising and staffing pose specific difficulties. The Block Development Officer (BDO) and his office look after the rural development programmes. BDO as well as his entire staff are drawn from the state govt. administration. While in administration, they have been dealing with mostly non-development works. In this case the decision making is centralised and the procedures have precedence over the results. The staff therefore carry along with them a burden of past approaches which inhibits their functioning as efficient development officers. Sound organisation structure calls for well-defined areas of work for the staff, effective delegation of authority and stimulation of creative thinking and initiative on the part of the staff.

There are only two ways of dealing with such a situation. One is appointing fresh staff especially for the office of BDO. The other option is to train and retrain the existing staff for enabling them to become capable of handling the new job requirement. Since the existing staff is already there, the first option is ruled out except in the case of new recruitment for filling up vacancies arising out of retirement of existing staff. The second option has therefore to be adopted. In this option, training should be aimed at improving the knowledge, skill, efficiency and most importantly the attitude. In the existing set up there is no recognition to the achievement, and this works as the most de-motivating factor as there is no distinction between the achievers and non-achievers. Even if there is some practical difficulty in awarding punishment for the non-achievers, there should be at least a jesture of reward for the achievers so that their morale is kept high. The reward need not be monetary only. It can even be in the form of acknowledging his good work.

Presently Human Resource Management (HRM) has acquired lot of importance. Basically HRM is concerned with the development of human resources to their full potential so that they may be able to give their best to the organisation.

Coordination and Control

Coordination

The organisation comprises number of employees working at various places and time frames. Most of the jobs are related to each other and the employees of one department also depend on other departments. In order to achieve desired results, there is a need to coordinate the efforts of all the concerned employees. Thus coordination is the management of the interdependence. It is required in all types of organisations including commercial organisations, public offices or NGOs whether working for profit or service. The deliberations in committees, group decision making, consultation process and involvement of concerned people in decision making process help in coordinating the efforts of various functionaries.

Several government and non-government organisations are connected with rural development programmes which are of multidisciplinary nature. Thus implementation of these programmes call for a serious nature of coordination because success of these programmes depend upon inputs from number of departments inside the organisation as well as outside. Many a times the coordination function in rural development programmes is faced with peculiar difficulties such as taking care of the ego of various functionaries. For example if a person in some connected department is, even if inadvertentfy, not invited for attending the meeting, there is every possibility that he may not extend the necessary cooperation. Thus this function puts heavy load on the executing department and the core staff responsible for implementation of the programme. The difficulties of such a nature can be greatly reduced by nomination of persons by name or designations who will represent the department concerned.

Another difficulty generally experienced is that the burden of coordination falls on very senior level persons who may be busy with many other works. Thus decisions remain pending till he finds time for calling a meeting or a group of concerned persons. In such a situation it is better to delegate the function to a person who can generally devote time for the programme.

Control

Management Expert—Anthony defines the control as, "the process by which managers assure that resources are obtained and used effectively and efficiently in the accomplishment of the organisations' objectives". Control function is closely linked with planning. It enables us to know if the results are following the way activities were planned. In case there is any deviation, it serves as a feedback so that the plan is reframed incorporating corrective action to be taken to achieve the objectives

What did take place	*What should take place*
Planning comprises:	*Control comprises:*
Policies	Cost Accounting
Programmes	Cost control
Procedures	Budgets
Management Information Systems	Management Reports

Thus control function enables to measure actual performance and guides the management in planning strategies to achieve the goals. Yearly budget wherein physical and financial parameters are given is the best tool of control and it is therefore widely used. The characteristics of Effective Control System are:

- It should promptly bring out deviations from the planned physical and financial progress.
- It should be forward looking so that the concerned authority can initiate corrective actions before substantial deviations occur.
- It should bring out deviations/exceptions particularly at those points where the performance significantly deviates from the planned progress.

Monitoring/Evaluation of Progress

Evaluation forms part of planning as well as control function. The objective of evaluation is to find out as to what extent the progress of the project/scheme has met the set targets. The criteria for evaluation should be related to the nature of the programme and its goals. There are various techniques used for evaluation of progress. In simple terms there are two, one of which is percentage progress system (PPS) and the other is bar chart system (BCS). Programe Evaluation and Review Technique (PERT) and Critical Path Method (CPM) are specialised techniques which requires knowledge as well as use of computers. The advantage of using PERT and CPM is that they show the interdependency between various activities and the path which is critical for reaching the goal. The difference between PERT and CPM is that while the former provides a way of computing the most probable time, the latter emphasises cost as well as time.

Reading of PERT requires familiarity and it requires a lot of work in updating the PERT. There has to be a cell dedicated for monitoring the progress. Therefore even in big projects it is used initially at the time of preparing project report and then used intermittently.

It is useful to mention in the remarks column of the bar chart about the reasons for the shortfall if any and the steps to be taken to make up the shortfall. Since interdependency of activities cannot be totally ignored, the use of PERT/CPM for selected macro activities and BCS in general offers the best option.

The Progress Bar Chart and a typical network, used in PERT & CPM, ideal tool for monitoring the progress of any rural programme/ Project/Scheme, are given at Fig. 6.1 and 6.2 respectively. In the case

Progress (bar chart) of Distribution of Surplus Land in District.......... during the Year......

As on 30th September..........

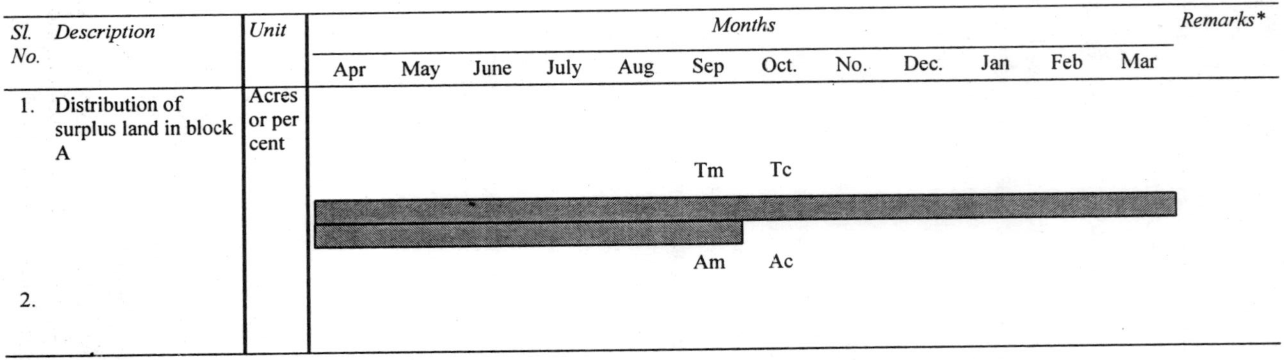

Tm : Target for the month

Am: Achievement during the month

Tc: Target (cummulative) upto the end of themonth

Ac: Achievement (cummulative) upto the end of the month

* Remarks (Col.): Reasons for shortfall and steps required to be taken to make up the shortfall.

Fig. 6.1: Typical Progress Bar Chart for Monitoring Progress of Rural Schemes

of network it is once again emphasised that it requires regular updating and use of computer, without which a network remains a showpiece.

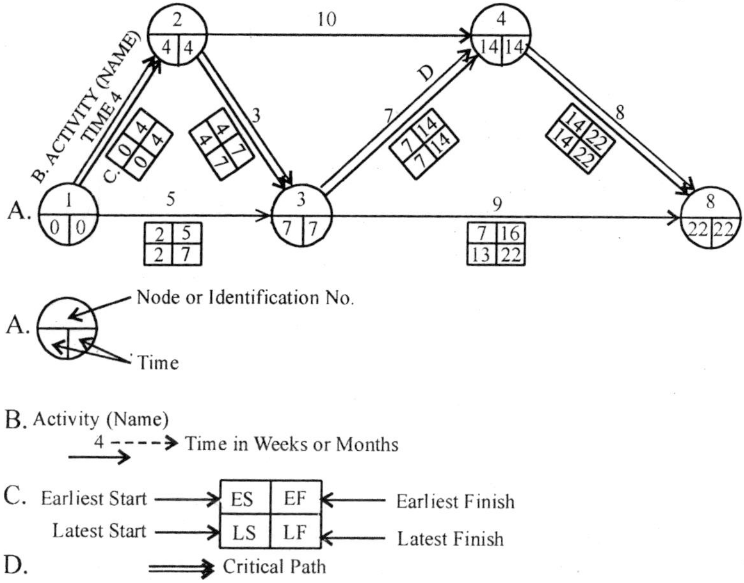

Fig. 6.2: Network Used in PERT and CPM for Monitoring Progress of Rural Schemes

If the programme is predominantly of social nature like creating awareness about the family welfare programme, there would be problems for quantification of the targets and therefore the evaluation criteria would have to be different. In this case the benefits are there but they are not tangible.

Monitoring of 20-Point (TPP) Programme

In rural development programmes, usually percentage progress system or bar chart system are used. Being simple, these are easier to follow. Table 6.1 shows all India performance of 20-point programme for the year 2002-03 and 2003-04. The monitoring of the programme is based upon Percentage Progress System. The programme was initiated in the year 1975 and restructured twice in 1982 and 1986.

The 20-Point Programme (20-PP or TPP) was conceived for monitoring of a number of schemes implemented by the Govt of

India and the State Governments with the objective of improving the quality of life of the people, especially those living below the poverty-line. It was meant to give a thrust to schemes relating to poverty alleviation, employment generation in rural areas, housing, education, family welfare and health, protection of environment and many other schemes having a bearing on the quality of life in the rural areas.

The present programme, known as TPP-86, comprises 119 items, out of which 54 are monitored on the basis of evaluatory criteria and 65 against pre-set physical targets. Out of the latter, 20 important items have been selected for monitoring on a monthly basis. The targets are fixed by the administrative Ministries at the Centre in consultation with the State Governments on the basis of budgetary outlays.

At the Central Govt. level, while the administrative Ministries monitor the specific schemes they are responsible for, the Ministry of Programme Implementation (MOPI) monitors the whole gamut of the schemes included in the programme. The MOPI brings out monthly and half-yearly progress report on the implementation of the programme. The monthly progress report covers achievements during each month and for the cumulative period against pre-set physical targets of selected 20 critical items, which have a direct bearing on the quality of life. The states are ranked every month based on their performance in respect of 12 key items out of the 20 items chosen for monthly monitoring.

Shortcomings in the Present System of Progress Monitoring and Evaluation

It is generally observed that the progress reports pertaining to rural development programmes are not updated regularly. There is also lack of authenticity of data being generated at the field level, major reason for this being the senior level officers feel good when the juniors present rosy picture of the progress. Needless to say often the figures are cooked up at the field level. As such no meaningful discussion can take place or decision taken on the basis of old and unauthentic data. The remedy for such practice is to give evaluation work to third party i.e. NGO. If the work is assigned to reputed NGO, such a system can succeed, but many a times NGOs in general are reluctant to give a critical report lest they loose the job next time.

It is pertinent to note that in the 20-Point Programme which is monitored very closely, one of the most important items namely basic education does not find place. Considering the great importance of basic education in rural area, this is simply inexplicable. Pre-set physical targets should be fixed for the *Sarva Shiksha Abhiyan* and they should very much form part of the 20-Point Programme.

Table 6.1: All India Performance of 20-Point Programme 2002-03 and 2003-04

Sl No	*Item*	*Unit*	*2002-03*			*2003-04*		
			Target	*Ach.*	*Per cent*	*Target*	*Ach.*	*Per cent*
1.	Sampoona Gramin Rozgar Yojana@	Lakh mandays	—	6407	—	—	7524	—
2.	Distributuion of surplus land	000 Acre	63	32	51	18	24	136
3.	Drinking water Prɔb (V/H)	000 Nos	64	41	63	111	41	37
4.	Community Health Centres (CHC)	Nos	267	33	12	575	148	26
5.	Primary Health Centres (PHC)	Nos	103	10	10	406	41	10
6.	Immunisation of children	Lakh No	252	232	92	254	274	108
7.	ICDS Blocks Operational (Cum)	Nos	5089	4903	96	5132	5263	103
8.	Anganwadis Operational (Cum)	000Nos	625	600	96	673	641	95
9.	SC Families Assisted	000Nos	2358	2305	98	2400	2559	107
10.	ST Families Assisted	000Nos	1138	1210	106	1117	1410	126
11.	Indira Awas Yojana (new houses)	000Nos	1314	1102	84	1485	1198	81
12.	EWS Houses Provided	000Nos	97	68	71	58	58	100
13.	LIG Houses	000Nos	27	16	57	7	13	187
14.	Slum Improvement (Popl. Cov.)	000Nos	5428	6345	117	3671	4996	136
15.	Tree Plantation	Lakh No	9171	8214	90	8622	7610	88
16.	Area to be covered	000Hect	1073	726	68	1072	844	79
17.	Villages Electrified	Nos	12363	3662	30	24011	5475	23
18.	Pump sets Energised	000Nos	242	337	139	217	232	107
19.	Improved Chullahs	000Nos	—	76	—	—	78	—
20.	Bio-gas Plants	000Nos	101	115	114	133	117	883

Source: India 2005 (GOI Publication)

@ No targets are fixed

V-Villages

H- Habitations

cum- cumulative

The existing mechanism of monitoring of the schemes is weak as the latest data is not generally available at the decision making level. Taking into account vast resources, both financial and physical being invested in these schemes, it is essential that their monitoring is of the same level as is practiced in the industry sector. With the phenomenal advances made in the Information Technology, this should not be difficult

20. Summing Up

To sum up. the most disturbing factors on rural front are:

Basic Education

i) There are 1.3 crore of children in the age group of 6-14 yet to be enrolled. The female literacy is just 47%.
ii) For every 100 children enrolled in class-I, only 47 reach class VIII. The dropout rate in the case of rural schools and that too of girls up to class-VIII is more than 53%.
iii) The quality of education in rural area is far too low compared to urban area.
iv) The absentee rate of teachers various between 14 to 31 percent.
v) Only 35 percent (as on 1st April 2003) of the children in primary classes have been covered so far with cooked mid-day meals in the schools.
vi) The financial allocation to education is 3.74 per cent (BE)/ 3 per cent (RE) of GDP in 2003-04 as against requirement of 6 percent. The financial allocation to basic education in rural area has all along been inadequate although attempt has been made to increase it by way of levying 2 per cent cess on direct taxes for this from 2004-05. The corresponding results are yet to be seen.

Primary Healthcare

i) The infant mortality rate per 1,000 live births is as high as 67 and percentage (under age 5) of underweight children for age is 47. The percentage of births attended by skilled health personnel is bare 43.

ii) Only 6 percent of households are having toilets and 21.9 percent of households are having latrines. On the other hand in sanitation programme, Govt is moving from high subsidy to low subsidy and instead proposes concentration on awareness campaign.

iii) About 10 percent of the total habitations are still without drinking water facility. Providing safe drinking water is still a bigger problem.

iv) The absentee rate of doctors and medical staff varies between 35 to 58 per cent as well as shortage of doctors and medical staff in rural health centers is to the tune of 36 per cent.

v) The financial allocation to health and family welfare is bare 0.9 per cent of GDP.

The financial allocation to primary healthcare in rural area is far too low.

Infrastructure

i) Only 60 per cent connectivity by road in rural area has been achieved.

ii) 16 per cent of the villages are yet to be provided with electricity.

iii) Efficient system of rural marketing is not in place.

Agriculture

i) Before 1996, the agricultural growth rate was about 3.4 per cent. But after this period the growth rate has been 1.7 per cent

ii) There is a downward trend in fhe share of agriculture in GDP and the sector now produces 21 per cent of GDP (2003-04).

iii) Foodgrain production is down by 18.2 per cent in 2002-03 and 2.8 per cent in 2004-05 w.r.t. base line production of 212 million tonnest in 2001-02 and 2003-04 as against fhe population increase of about 2 per cent every year. So also the share of exports of agricultural products in the total exports has gone down to 14.7 per cent.

iv) There is a downward trend in the outlay on agriculture and allied activities as a percentage of total plan outlay and the same stands at 10.6 per cent during X Plan.
v) The net area under irrigation has gone down from 56.76 mha in 1999-2000 to 54.68 mha in 2000-01.
vi) There is a downward trend in the share of electricity sold to agriculture and the same stands at bare 24.88 per cent.
vii) The benefit of rural credit is not reaching small and marginal farmers as compared to big farmers
viii) The share of cooperative banks in the total rural credit is dwindling fast.and the same is now 28 per cent.

General

i) 27.09 per cent of rural population falls below poverty line.
ii) About 20 per cent of the amount allocated to the schemes actually reaches the real intended beneficiaries. The rest goes to payment of salaries of staff or administrative expenditure, inefficiencies in implementation of schemes and leakages at various levels.
iii) In spite of the fact that it is difficult to provide even minimum respectable living standard to every household for a long time to come, there is unwillingness to adopt two-child norm which is necessary to control the present rate of population growth, at least for the period such a norm is essential. On the other hand China is still pursuing one-child norm. At the present rate, Indian's population is expected to overtake that of China in 2030 and become the largest populated country in the world.
iv) In Human Development Index (UNDP HDR-2004), China has improved its rank from 104 in 2001 to 94 in 2002, whereas India is static at the lower end at 127 in the comity of 177 nations.

On the plus side

i) India ranks first in the milk production in the world.
ii) India accounts for one sixth of the world's cattle and more than half of the world's buffalo population.
iii) India is the largest producer of jute and allied fibres.

iv) India is fourth largest fish producing country in the world with 6 million metric tons of fish production annually.

v) India's production of foodgrains has gone beyond its need and it is in a position to export the foodgrains. It is a different matter that more than 200 million remain undernourished, primarily due to failure of Public Distribution System.

The above points have been substantiated with data and, statistics at respective places in the book along with strategies required to be adopted to achieve the desired developmental goals.

There is no doubt that there is growing awareness on the part of public as well as people's representatives about the dimension of the problems being faced by the rural population and this is reflected in the bold steps taken up by the Governments like starting flagship Programme–Sarva Shiksha Abhiyan and Bharat Uday in 2002, increased allocation to elementary education even by levy of cess, enhancing agricultural credit, setting up National Rural Health Mission, framing National Employment Guarantee Scheme in 2004/2005 etc.

In his address to Parliament on 25th February 2005, the President outlined Govt's vision to build India and called it 'Bharat Nirman' As announced by the Union Govt. on 16.05.2005 Bharat Nirman has been conceived as a programme to be implemented with an investment of Rs. 1,74,000 crore over a period of 4 years, to build infrastructure, especially in rural India. The specific goals are:

- to bring an additional one crore hectares under assured irrigation;
- to connect all villages that have population of 1000 (or 500 in hilly/tribal areas) with a road;
- to construct 60 lakhs additional houses for the poor;
- to provide drinking water to 74,000 habitations that are uncovered;
- to provide electricity to remaining 1,25,000 villages through Rajiv Gandhi Gramin Vidyutikaran Yojana (RGGVY) and offer electricity connection to 2.3 crore households; and
- to give telephone connectivity to remaining 66,822 villages.

A 'New Deal' as a part of 'Bharat Nirman' thus envisages higher investment in agriculture, assured irrigation facility and providing food and nutritional security while at the same time augmenting farm incomes and employment as a sequel to launching of the second green revolution.

While the programmes taken up by various Governments deserve utmost appreciation, it is more important as to how far the intentions are translated into results at the ground level. In this context the progress achieved in the rural programmes, so far, leaves much to be desired.

21. Concluding Remarks

Rural Development is a complex phenomenon as it comprises number of important activities to be taken up simultaneously. As suggested in the framework of Human Development Index designed by United Nations Development Programme for measuring the progress of a country, when translated to rural development, progress in the areas of basic education, primary healthcare and raising rural income levels is most crucial. The desired progress in the aforesaid areas can only be achieved through development of rural infrastructure, allocation of adequate funds for rural development schemes and efficient public delivery system. In this context the commitment of political class, accountability of bureaucracy for proper implementation of various rural development schemes and active participation in the schemes of the community for whose benefit they are designed play most crucial role in achieving the desired goals.

In nutshell, efforts of multitude of organisations and their people who are involved in the whole development process have to be so integrated that they produce strong synergy to accomplish the task. While the efforts made and the progress achieved so far are commendable, they leave much to be desired and fall short of the genuine expectations of the rural population. It is just impossible to achieve the desired goals without adopting certain strategies as required. These strategies are elaborated in great detail at the end of each chapter. It is hoped that those who are concerned directly or indirectly with rural development programmes can benefit from the recommendations made in this direction from place to place in this book.

Bibliography

Aggarwal, Y.P., 1999. *Access and Retention in District Education Programme Districts'.* A National Overview, Paper presented. National Council of Educational Research and Training, New Delhi.

Aggrawal, A.N. and S.P.Singh. 1975. *The Economics of Under Development,* Oxford University Press, New Delhi.

Agriculture in Maharashtra, Yesterday, Today and Tomorrow: Krishi Saptak Programme, Govt. of Maharashtra, Pune.

Aikara, Jacob, 1997. *Learner Achievement in Primary School,* Unit for Research in the Sociology of Education, Tata Institute of Social Sciences, Mumbai.

Anandalakshmy, S. and Jain Sharda, 1997, Shikshakarmi: *A Paradigm Shift in Primary Education,* Sandhan Research Centre, Jaipur.

Anitha, B.K. 2000. *Village, Caste and Education,* Rawat Publications, Jaipur.

Annual Report 2002-03. Ministry of Health and Family Welfare, New Delhi.

Athreya, V. and Sheelarani Chunkath. 1996. *Literacy and Empowerment,* Sage Publications, New Delhi.

Aurora, D. 1977. *Drought Prone Area: Some Development Issues,* Institute of Public Administration, New Delhi.

Ashokvardhan, C. (IAS), *Socio-Economic Profile of Rural India* (Vol. 2: North East India), Concept Publishing Company, New Delhi.

Basu, D.N. and G.S.Guha. 1996. *Agro-Climatic Regional Planning in India, Vol. I: Concepts and Application,* Concept Publishing Company, New Delhi.

Basu, D.N. and G.S.Guha. 1996. *Agro-Climatic Regional Planning in India, Vol. II: Themes and Cases,* Concept Publishing Company, New Delhi.

Basu, Durga Das, 1994. *Introduction to the Constitution of India,* Prentice-Hall of India, New Delhi.

Bhatia, B.M. 1977. *Poverty, Agriculture and Economic Growth,* Vikas Publishing House Pvt. Ltd., New Delhi.

Bulletin on Rural Health Statistics in India. June 2000, Rural Health Division, DGHS, Ministry of Health and Family Welfare, New Delhi.

Bansil, P.C., *Water Management in India*, Concept Publishing Company, New Delhi.

Census of India 2001. Registrar General and Census Commissioner, India, New Delhi.

Central Budgets, placed in Parliament; *Economic Times*, published from Mumbai.

Chandra, Bipin. 1989. *India's Struggle for Independence*, Penguin Books India Pvt. Ltd., New Delhi.

Chattopadhyay, B.C. 1985. *Rural Development Planning in India,* S. Chand & Co. Ltd., New Delhi.

Chitnis, Suma, 1981. *A Long Way to Go,* Allied Publishers, New Dehi.

Dandekar, V.M. and Nilkanth Rath. 1971. *Poverty in India,* Indian School of Political Economy, Pune.

Dantwala, M.L. 1986. *Agrarian Structure and Agrarian Relations in India,* in *Indian Agricultural Development since Indenendence,* Indian Society of Agricultural Economics, Oxford and IBH Publishing Co. Pvt. Ltd., New Delhi.

Desai, A.R., 1978. *Rural Sociology in India,* Popular Prakashan, Mumbai.

Desai, D.K., 1983. *Management in Rural Development,* Oxford and IBH, New Delhi.

Desai, S.S.M. 1986. *Rural Banking in India,* Himalaya Publishing House, Mumbai.

Desai, Vasant. 1983. *A Study of Rural Economics,* Himalaya Publishing House, Mumbai.

Deshpande, D.V., M.K. Mudgal and K.K. Gupta. 1996. *Regional Rural Banks at Cross Roads,* Working Paper, Bankers Institute of Rural Development, Lucknow.

Doshi, S.L. and Jain, P.C. 2002. *Rural Sociology,* Published by Rawat Publications, Jaipur and New Delhi.

Das, Purnendu Sekhar, *Decentralized Planning and Participatory Rural Development*, Concept Publishing Company, New Delhi.

Economic Times, published from Mumbai.

Gangrade, K.D. 1997. *Power to Powerless: A Silent Revolution through Panchayat Pa/System,* in journal of Rural Development. 6(4).

Gopalaswamy, T.P. 1998. *Rural Marketing-Environment Problems and Strategies,* Wheeler Publishing Co., New Delhi.

Gupta, D.N. (IAS), *Decentralisation: Need for Reforms*, Concept Publishing Company, New Delhi.

Hanumanta, Rao C.H. 1965. *Agricultural Production and Returns in India,* Institute of Economic Growth, Asia Publishing House, Delhi.

Hingorani, Anand T. and Hingorani, Ganga A. 1985. *The Encyclopaedia of Gandhian Thoughts,* AICC Publications Department, New Delhi.

Hitavada, *Newspaper published from Nagpur.*

Shylendra, H.S., *Diversification and Sustainable Rural Livelihood,* Concept Publishing Company, New Delhi.

India 2005, A Reference Annual, compiled and edited by Research, Reference and Training Division, Publications Division, Ministry of Information and Broadcasting, Govt. of India, New Delhi.

India Education Report 2002, National Institute of Educational Planning and Administration and United Nations Educational, Scientific and Cultural Organisation, Oxford University Press, New Delhi.

India Sustaining Reform, Reducing Poverty. A World Bank Development Policy Review 2003, OXFORD University Press, New Delhi.

Jain, S.C. 1985. *Problems of Agricultural Development in India,* Kitab Mahal, Allahabad.

Jain, S.C., Indigenous Resources for Rural Development, Concept Publishing Company, New Delhi.

Jana, Anil Kumar, Administering District Plans in India: Issues, Constraints & Choices, Concept Publishing Company, New Delhi.

Kaul, Rekha. 2001. Accessing Primary Education: *Going Beyond the Classroom,* Economic and Political Weekly, vol. 36.

Krishi Margadarshika. 1999. Dr. Panjabrao Krishi Vidyapeeth, Akola.

Kumar, Krishna, 1989. *Social Character of Learning,* Sage Publications, New Delhi.

Kumar, Nalini. 1997. *Operation Flood: Literature Review and Reconcilation,* Occasional Publication, Institute of Rural Management.

Majumdar, D.N., 1958. *Caste and Communication in an Indian Village,* Asia Publishing House, Mumbai.

Mascarenhas, 1988. *A Strategy for Rural Development: Dairy Cooperatives in India,* Sage Publications, New Delhi.

Mehta, Shiv R. 1984. *Rural Development Policies and Programmes,* Sage Publications, New Delhi.

Munjal, S. 1999. Raj Publishing House, New Delhi and Jaipur.

Myrdal, Gunnar 1968. *Asian Drama-on Enquiry into the Poverty of Nations,* Vol. II, The Penguin Press, London.

Mukherjee, Amitava, *Participatory Rural Appraisal: Methods and Appications in Rural Planning* (Essays in Honour of Robert Chambers), Concept Publishing Company, New Delhi.

Mukhernee, Amitava, *Participatory Learning and Action and Monitoring and Evaluation and Participatory Monitoring and Evaluation* (Essays in Honour of Robert Chambers), Concept Publishing Company, New Delhi.

National Health Policy 2002, Department of Health, Ministry of Health and Family Welfare, New Delhi.

Nehru, Pt. Jawaharlal. 1998 (18th Impression). *Discovery of India,* Teen Murti House, New Delhi.

Pal, Suresh and Alka Singh. 1997. *Agricultural Research and Extension in India:Institutional Structure and Investments,* Policy Paper, National Centre for Agricultural Economics and Policy Research, New Delhi.

Pareek, Uday. 1982. *Education and Rural Development in Asia,* Oxford and BBH Publications, New Delhi.

Park, K. 2005. *Preventive and Social Medicine,* Banarasidas Bhanot Publishers, Jabalpur.

Parthasarthy, G. 1979. *Land Reforms and Changing Agrarian Structure,* in *Agricultural Development of India-Policies and Problems,* (ed.) C.H. Shah. Orient Longman, Mumbai.

Prasad, Lallan and Gulshan, S.S. 2001. *Management: Principles and Practices, & Chand & Company,* New Delhi.

Prasad, Govind and Mishra, Anil Dutta, (Eds.), ????. *Globalisation: Myth and Reality*, Concept Publishing Company, New Delhi.

Palanithuri, G., *Rural Transformation and Peoples Entitlements*, Concept Publishing Company, New Delhi.

Publishing House, Mumbai.

Rao, Usha. N.J.I 983. *Women in a Developing Society,* Asish Publishers, New Delhi.

Rao, R.M. Mohan, Growth and Rural Transformation: A Comparative Study of the Republic of Korea and India, Concept Publishing Company, New Delhi.

Reddy, K.Venkata, 2001. *Agriculture and Rural Development: A Gandhian Perspective,* Himalaya Publishing House, Mumbai.

Report on Village Facilities 2002, NSS 58 Round. National Sample Survey Organisation, Ministry of Statistics & Programme Implementation, GOI, New Delhi.

Rural Development Report 1999. Centre on Integrated Rural Development for Asia and the Pacific (CIRDAP).

Rural Development Statistics 2002-03. National Institute of Rural Development, Hyderabad, Ministry of Rural Development, GOI.

Shah, C.H. 1979. *Agricultural Development in India-Policy and Problems,* Orient Longman, Mumbai.

Sharma, K.L., 1997. *Social Stratification in India: Issues and Themes,* Sage Publications, New Delhi.

Sharma, V.K. and Satendra (IFS), *Sustainable Rural Development for Disaster Mitigation*, Concept Publishing Company, New Delhi.

Singh, Kartar. 1999. *Rural Development: Principles, Policies and Management, Sage Publications,* New Delhi.

Singh, Prabhakar. 1982. *Community Development Programme in India,* Deep and Deep Publications, New Delhi.

Singh, R.P. 1998. *Implications of GATT/WTO on Agriculture and Rural Development,* National Institute of Rural Development, Hyderabad.

Singh, Yogendra, 1985. *A Survey of Research in Sociology and Social Anthropology (1969-19790),* Satvaham Publications, New Delhi.

Singh, Yogendra, 1986. *Indian Sociology,* Sage Publications, New Delhi.

Srinivas, M.N. 1978. *India's Villages,* Media Promoters and Publishers Pvt. Ltd., Mumbai.

Srinivas, M.N. 1982. *Social Structure,* Hindustan Publishing Corporation, Delhi.

Statistical Abstract. India 2003, Central Statistical Organisation, Ministry of Statistics and Programme Implementation, Govt. of India, New Delhi.

Sushila Mehta, 1980. *A Study of Rural Sociology,* S. Chand & Co. Ltd., New Delhi.

Swaminathan, Mina, 1999, The First Five Years: *A Critical Perspective on Early Childhood Care and Education in India,* Sage Publications, New Delhi.

The Consitution of India 2003. Ministry of Law and Justice, Govt. of India, New Delhi.

Tilak, J.B.G., 1987. *Economics of Inequality in Education,* Sage Publications, for Institute of Economic Growth, New Delhi.

Times of India, Newspaper published from Mumbai.

UNDP Human Development Report 2004. Cultural Liberty in Today's Diverse World. Published for the United Nations

Development Programme (UNDP), OXFORD University Press, New Delhi.

Vaidyanathan, A. and Nair, P.R.Gopinathan. *Elementary Education in Rural India: A Grassroots View,* Sage Publications, New Delhi.

WHO 1981 Global Strategy for Health for All by the Year 2000.

World Bank 1997. Primary Education in India, Allied Publishers, New Delhi.

Index